Contents

Unit 1   Properties and Shapes

Section 1   Presentation

1. Look and read:

Lines

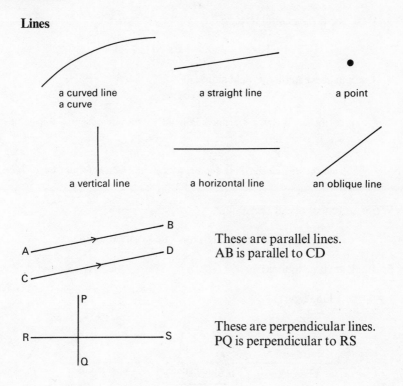

a curved line
a curve

a straight line

a point

a vertical line

a horizontal line

an oblique line

These are parallel lines.
AB is parallel to CD

These are perpendicular lines.
PQ is perpendicular to RS

Now describe the following lines:

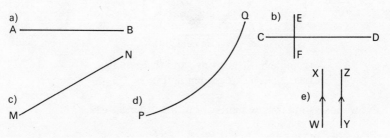

a)
A ———————— B

Q   b)   E
C ———————— D
          F

c)
M

N

d)
P

X   Z

e)
W   Y

2. Look at these Greek letters:

∝	π	δ	β	θ	φ
alpha	pi	delta	beta	theta	phi

1

Now answer these questions:

    a) Which letter has two parallel vertical lines and one horizontal line?

    b) Which letter has one curved line and a horizontal line?

    c) Which letter has one curved line and a vertical line?

    d) Which letters have one curved line?

    e) Which letter has one straight line and two curved lines?

3. Describe these capital letters:

    a) H   b) K   c) O   d) M   e) U

4. Look at these mathematical signs:

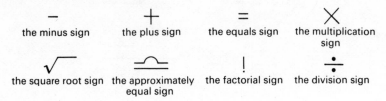

Write sentences about these signs:

Example: The minus sign has one horizontal line.

5. Look at these figures:

Plane figures

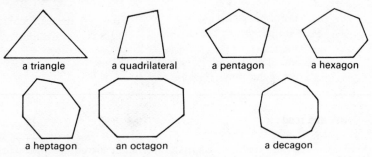

A triangle has three sides and three angles.
A triangle is a three-sided figure.

Now make similar statements about the other figures.

6. Look and read:

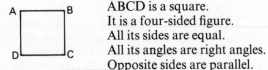

ABCD is a square.
It is a four-sided figure.
All its sides are equal.
All its angles are right angles.
Opposite sides are parallel.

Now describe these figures:

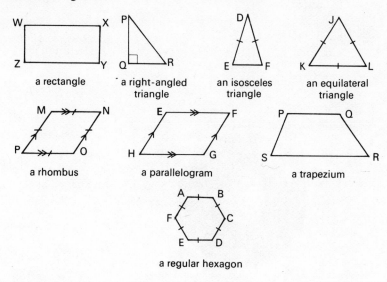

a rectangle   a right-angled triangle   an isosceles triangle   an equilateral triangle

a rhombus   a parallelogram   a trapezium

a regular hexagon

7. Look at these examples:

- A square is a kind of rectangle, but not all rectangles are squares.
- A right-angled triangle is a kind of triangle, but not all triangles are right-angled.

Now make similar statements about the following:

a) rhombus; parallelogram
b) parallelogram; trapezium
c) isosceles triangle; triangle
d) isosceles triangle; equilateral triangle
e) rectangle; plane figure

8. Look and read:

- All the sides of an equilateral triangle are equal, and so are those of a rhombus.
- Opposite sides of a parallelogram are parallel and so are those of a regular hexagon.

Now complete the following sentences:

a) Opposite sides of a rectangle are equal, and so are ......
b) All the angles of a square are right angles, and so are ......
c) All the sides of a regular hexagon are equal, and so are ......
d) All the sides of an equilateral triangle ......
e) All the angles of an equilateral triangle ......

3

9. Look at these examples:

- The cover of this book is shaped like a rectangle. It is rectangular.

- A record is shaped like a circle. It is circular.

Now use this table to write sentences about the objects below:

circle	– circular	pentagon	– pentagonal
semi-circle	– semi-circular	hexagon	– hexagonal
triangle	– triangular	octagon	– octagonal
rectangle	– rectangular	heptagon	– heptagonal
square	– square		etc.

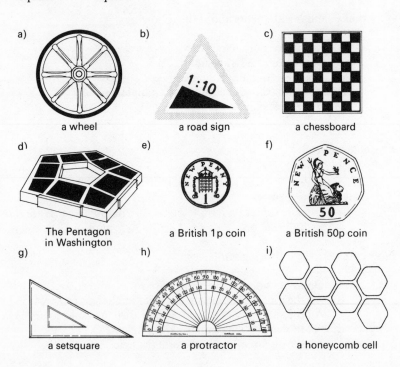

a) a wheel

b) a road sign

c) a chessboard

d) The Pentagon in Washington

e) a British 1p coin

f) a British 50p coin

g) a setsquare

h) a protractor

i) a honeycomb cell

4

Section 2   Development

10. Look and read:

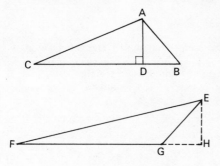

AD is an altitude of the triangle.
BC is the base.
What is the area of △ ABC?

AD is equal to EH and BC is equal to FG.
The altitudes of the two triangles are equal and so are the bases.
Therefore the areas are the same.
△ ABC has the same area as △ EFG.

Now describe the following pairs of triangles:

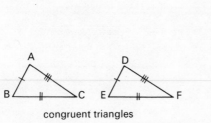

congruent triangles

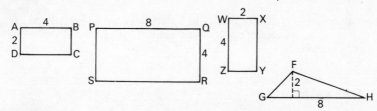

similar triangles

11. Compare these figures, saying whether they are congruent, similar or have the same area:

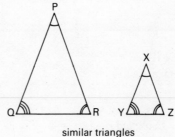

12. Look and read:

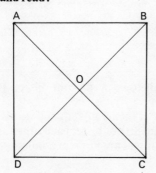

In square ABCD, the diagonals intersect at 0.

OA = OC   and   OD = OB.   The diagonals bisect each other.

$\widehat{AOB} = 90°$. The diagonals intersect at right angles.

$\widehat{DAO} = \widehat{BAO}$. The diagonals bisect the angles.

5

AC = BD. The diagonals are equal.

AB = DC. The opposite sides are equal.

AB//DC. The opposite sides are parallel.

Now complete these tables:

	Opposite sides equal	Opposite sides parallel	Diagonals bisect each other	Diagonals bisect angles	Diagonals intersect at right angles	Diagonals equal
Square	✓	✓	✓	✓	✓	✓
Rhombus						
Parallelogram						
Trapezium						
Regular hexagon						

	Angles equal	Sides equal	Areas equal
Congruent figures			
Similar figures			

13. Look and read:

The opposite sides of a square are equal and parallel. The diagonals bisect each other at right angles and also bisect the angles. Also, the diagonals are equal.

Now write similar paragraphs about a rhombus, a parallelogram, a trapezium, a regular hexagon, and congruent and similar figures, using the information in the above table.

14. Look and read:

The diagonals of a rhombus are not equal unless the rhombus is a square.

Now complete the following sentences:

a) The diagonals of a parallelogram do not bisect the angles unless......

b) The diagonals of a parallelogram are not equal unless......

c) The diagonals of a trapezium do not bisect each other unless......

d) The sides of two similar triangles are not equal unless......

Section 3 Reading

15. Read this:

The circle

A circle is a plane geometric figure. The side of a circle is called the circumference. All the points on the circumference of a circle are equidistant from the centre. A straight line which is drawn from the centre of a circle to its circumference is called a radius. All the radii of a circle are equal. The area between two radii is called a sector. A straight line drawn between one part of the circumference and another is known as a chord. A chord separates a circle into two segments and the circumference into two arcs. A chord which passes through the centre of a circle is called a diameter.

Name the following parts of the diagram:

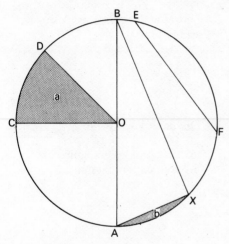

a) Shaded area *a*
b) Shaded area *b*
c) EF
d) XF
e) AB
f) OC and OD
g) O

Now answer these questions:

h) What are the properties of any triangle ABX with AB as its base and X as a point on the circumference? (see diagram).
i) When is a sector also a segment?

16. Say whether the following statements are true or false. Correct the false statements.

a) A chord is a curved line.
b) The radius of a circle is half the length of its diameter.
c) A closed curve where all points on the curve are equidistant from the centre is called a circle.
d) A sector has three sides – two chords and an arc.

7

Section 4   Listening

17. Listen to the passage and draw the diagram. Then answer the following questions:

   a) Name three figures in the diagram which have the same area as ABCD.

   b) How many right-angled triangles are there in the diagram?

   c) How many pairs of non-congruent similar triangles are there in the diagram?

   d) How many trapeziums are there in the diagram?

   e) How many quadrilaterals which are neither parallelograms nor trapeziums are there in the diagram?

   f) Which triangle has the same area as $\triangle$ ACD + $\triangle$ DGF?

18. PUZZLE

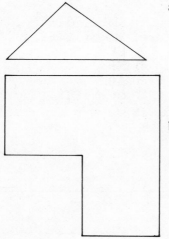

   a) Divide this triangle into three figures with equal areas, using two straight lines.

   b) Divide this figure into:
      i) two congruent figures
     ii) three congruent figures
    iii) four congruent figures

Unit 2   Location

Section 1   Presentation

1. Look at this:

Angles

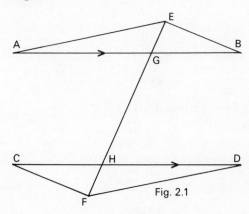

Fig. 2.1

Now describe Figure 2.1.

   (**Note:** EF is called a transverse line)

2. Look and read:

$A\widehat{G}H = E\widehat{G}B$	They are *vertically opposite* angles. (Vert. opp. $\angle$s)
$A\widehat{G}H = C\widehat{H}F$	They are *corresponding* angles. (Corr. $\angle$s)
$A\widehat{G}H = G\widehat{H}D$	They are *alternate* angles. (Alt. $\angle$s)
$A\widehat{G}H + A\widehat{G}E = 180°$	They are *adjacent angles on a straight line.* (Adj. $\angle$s)
$A\widehat{G}H + C\widehat{H}G = 180°$	They are *interior angles on the same side of the transverse line.* (Int. $\angle$s)
$A\widehat{G}H = E\widehat{A}G + A\widehat{E}G$	The *exterior angle* of a triangle equals the sum of the *interior opposite angles.* (Ext. $\angle$ of $\triangle$)

Now make similar statements about $E\widehat{G}B$:

   Find other angles in the diagram which are equal and say why.
   If EB is equal and parallel to CF, compare $\triangle$CHF and $\triangle$EGB.
   Give reasons for what you say.

3. Look and read:

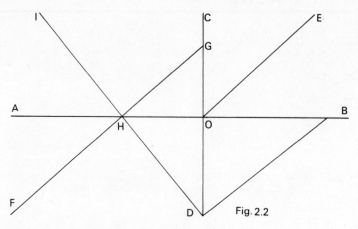

Fig. 2.2

In Figure 2.2, AB is perpendicular to CD

OE bisects $C\widehat{O}B$

DI bisects $O\widehat{H}F$

FG//OE//DB

Look at these examples:

OE bisects $C\widehat{O}B$ (Given)

$\therefore C\widehat{O}E = 45°$

$O\widehat{D}B = C\widehat{O}E = 45°$ (Corr.∠s)

Now complete the following statements in the same way:

a) OE bisects $C\widehat{O}B$ (_____) $\therefore E\widehat{O}B$ ......

b) $O\widehat{B}D = E\widehat{O}B$ ......

c) $A\widehat{H}F = O\widehat{B}D$ ......

d) $O\widehat{H}F + A\widehat{H}F$ ......

e) $\therefore O\widehat{H}F =$ _____

f) DI bisects $O\widehat{H}F$ _____

g) $\therefore A\widehat{H}I = G\widehat{H}I =$ _____

h) $O\widehat{H}D = A\widehat{H}I$ ......

i) $A\widehat{O}D =$ _____

j) $\therefore O\widehat{D}H =$ ......

4. Answer these questions about Figure 2.2:

a) Which triangles are similar?

b) How many isosceles triangles are there in the diagram?

c) Produce EO to meet DH at J. Describe △OHJ.

5. Look and read:

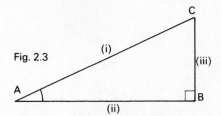

Fig. 2.3

In a right-angled triangle, the side opposite the right angle is called the hypotenuse.

a) Complete this statement of Pythagoras's Theorem.
The square of the ...... is equal to the sum of the ......
b) Label the three sides of △ABC in Figure 2.3, in relation to $\widehat{CAB}$, using the words *hypotenuse*, *opposite* and *adjacent*.

Complete the following:

c) $\text{tangent} = \dfrac{\text{opposite}}{\text{adjacent}}$ d) $\tan CAB =$

e) $\text{sine} =$ f) $\sin CAB =$

g) $\text{cosine} =$ h) $\cos CAB =$

6. Look and read:

Inscribed and circumscribed figures

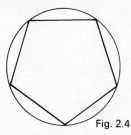

Fig. 2.4

Figure 2.4 shows a circle *circumscribed about* a pentagon. Each vertex of the pentagon lies on the circumference of the circle.

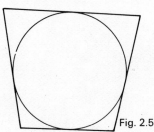

Fig. 2.5

Figure 2.5 shows a circle *inscribed in* a quadrilateral. Each side of the quadrilateral is a *tangent to* the circle.

Write correct sentences from the table:

A circle can be inscribed in	a quadrilateral	in all cases.
A circle can be circumscribed about	a triangle	
	a trapezium	
A circle cannot be inscribed in	a rhombus	unless the sums of the opposite sides are equal.
	a rectangle	unless the sum of the opposite angles is 180°.
A circle cannot be circumscribed about	a square	unless the nonparallel sides are equal.

Section 2   Development

7. Look and read:

Simple matrices

Figure 2.6 shows a simple 5 × 5 matrix.
> 5 is in the top left-hand corner of the matrix.
> 15 is immediately below 33.
> 34 is immediately to the left of 16.
> 6 is immediately above 24.

$$\begin{pmatrix} 5 & 9 & 33 & 1 & 21 \\ 17 & 23 & 15 & 4 & 27 \\ 34 & 16 & 7 & 3 & 6 \\ 2 & 12 & 18 & 8 & 24 \\ 11 & 14 & 19 & 4 & 13 \end{pmatrix}$$

Fig. 2.6

Describe the position of the following numbers:

21;   6;   13;   7;   11;   23;   9

8. Look and read:

33 is the first element of the third column of the matrix. (Figure 2.6)
34 is the first element of the third row of the matrix.
5 and 33 are the first elements of the first and third columns respectively.
27 and 6 are the last elements of the second and third rows respectively.

12

Describe the position of these numbers in the same way:

2;   23;   14;   15 and 7;   8 and 4;   8 and 24.

9. Look and read:

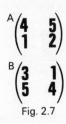

A $\begin{pmatrix} 4 & 5 \\ 1 & 2 \end{pmatrix}$

B $\begin{pmatrix} 3 & 1 \\ 5 & 4 \end{pmatrix}$

Fig. 2.7

4 in matrix A corresponds to 3 in matrix B.
a) What are the other corresponding elements in the two matrices?
b) Explain how to add two matrices.
c) Explain how to multiply two matrices.

Section 3   Reading

10. Read this:

Angles
In plane geometry an angle is a figure which is formed by two straight lines which meet at a point. The lines of an angle are called the sides. The point where they meet is called the vertex.

When the sides of an angle are perpendicular to each other, they form a right angle. A right angle has ninety degrees. An angle of less than 90° is an acute angle, and an angle of more than 90° but less than 180° is an obtuse angle. An angle of more than 180° is a reflex angle.

Triangles are often named according to their angles. A right-angled triangle has one right angle and two acute angles. An obtuse triangle has one obtuse and two acute angles. An acute triangle has three acute angles.

Now answer the questions:

What kind of angle does a clock make at

a)   two o'clock?
b)   three o'clock?
c)   four o'clock?
d)   twenty to ten?
e)   twelve minutes past seven?
f)   twenty-nine minutes past twelve?

11. Name the kinds of angle shown in Figure 2.8:

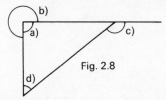

Fig. 2.8

12. **Say whether the following statements are true or false. Correct the false statements.**

   a) The exterior angle of a triangle is always obtuse.
   b) Only two angles of a triangle can be acute.
   c) The smallest angle of a triangle is opposite the shortest side.
   d) The point where the sides of an angle meet is called the vertex.
   e) A triangle with two obtuse angles is called an obtuse triangle.

Section 4   Listening

13. **Listen to the passage and draw the diagram, then answer the questions:**

   a) Find the values of the following angles, giving reasons:

   i) $A\widehat{E}B$
   ii) $O\widehat{E}K$
   iii) $E\widehat{B}A$
   iv) $B\widehat{E}K$
   v) $H\widehat{G}B$

   b) How many pairs of congruent triangles are there in the diagram?
   c) How many pairs of noncongruent similar triangles are there in the diagram?

14. **Say whether the following statements are true or false. Correct the false statements.**

   Join AL and BM

   a) ABDC is an isosceles trapezium
   b) BD is parallel to LA
   c) BX, a fourth side to the square EOBX, is a tangent to the circle
   d) The pentagon AEBML is regular
   e) The sum of $E\widehat{A}M$ and $E\widehat{B}M$ is two right angles
   f) The hexagon AEBMNL has the same area as the hexagon AFGBML

15. PUZZLE

How many triangles are there in this diagram?

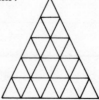

Unit 3   Structure

Section 1   Presentation

1. Look and read:

Operations on numbers

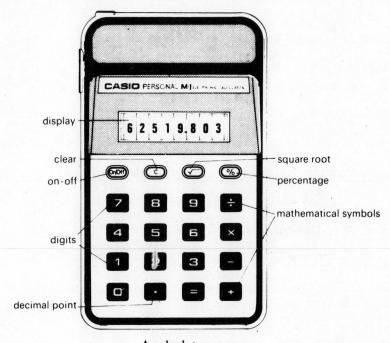

A calculator

- This calculator *consists of* a plastic case, electronic circuits, a display and twenty keys.
- The plastic case *contains* batteries.
- The keys *include* ten digit keys, one decimal point key, and so on.
- The keys are *arranged in* four columns and five rows.
- The middle row *comprises* three digits and the multiplication sign.
- The number in the display *contains* eight digits.

Now ask and answer questions like the following examples:

What does the middle row consist of?
Which keys does the middle row comprise?
How are the digit keys arranged?
How many digits does the number 23 462 contain?

15

2. Look and read:

Arithmetical operations on numbers include addition, subtraction, division and multiplication.

One number may be added to another. The result is called the sum. The sum of 9 and 14 is 23.

Make similar statements using these words:

a) subtracted/difference
b) multiplied/product
c) divided/quotient

3. Look and read:

An integer is even if it is divisible by 2.
An integer is odd if it is not divisible by 2.
An integer is divisible by 3 if the sum of its digits is divisible by 3.

Now make similar statements about the divisibility of integers by:

a) 10 b) 9 c) 4 d) 8 e) 5 f) 6 g) 11

4. Look at this set of numbers:

2, 3, 5, 7, 11, 13, 17, 19, 23 ......

a) Can you continue this set? It is made up of prime numbers.
b) What is a prime number?

Section 2   Development

5. Look and read:

Fractions

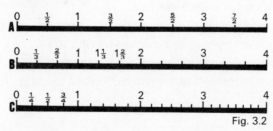

Fig. 3.2

a) In A, each unit is divided into halves.
b) In B, what is each unit divided into?
c) What is the first unit divided into in C?
d) Ask and answer similar questions about the other units in C.

6. Read this:

A number such as $\frac{3}{5}$ is called a fraction. A fraction comprises two parts, the denominator and the numerator. The denominator is the number below the line.

a) What is the numerator?
b) What are the numerator and denominator *separated by*?

7. Read this:

If the numerator is less than the denominator, the fraction is known as a proper fraction.
If the denominator is less than the numerator, the fraction is known as an improper fraction.
In the fraction $\frac{102}{153}$ both the denominator and the numerator may be divided by the same number (51) to give $\frac{2}{3}$.

Make similar sentences about these fractions:

a) $\frac{10}{16}$ b) $\frac{24}{80}$ c) $\frac{17}{102}$ d) $\frac{342}{360}$ e) $\frac{243}{405}$

This is called *cancelling* or *reducing* the fraction. Can the following fractions be reduced?

f) $\frac{28}{70}$ g) $\frac{81}{450}$ h) $\frac{40}{64}$ i) $\frac{15}{495}$ j) $\frac{41}{105}$

8. Look and read:

- Both 12 and 18 are divisible by 6.
- 12 and 18 are both divisible by 6.
- Neither 12 nor 18 is divisible by 5.
- 18 is divisible by 9, whereas 12 is not (divisible by 9).
- 18 is divisible by 9. 12, on the other hand, is not (divisible by 9).

Now make similar sentences about the following pairs of numbers:

a) 10, 20   b) 14, 21   c) 118, 354

9. Look and read:

Any integer may be *represented* as the product of prime numbers.
For example, $150 = 2 \cdot 3 \cdot 5^2$.
This is known as factorising a number.
20 can be factorised into $2^2 \cdot 5$.

Make similar statements about these numbers:

a) 16   b) 24   c) 36   d) 370

10. Read this:

> 150 and 20 both have the factors 2 and 5. 150 and 20 have the factors 2 and 5 *in common*. The highest common factor (H.C.F.) of 150 and 20 is $2 \cdot 5 = 10$. The H.C.F. is also known as the greatest common divisor. The lowest common multiple (L.C.M.) of 20 and 150 is $2^2 \cdot 3 \cdot 5^2 = 300$. The L.C.M. is also known as the least common multiple.

Ask and answer questions about the highest common factors and lowest common multiples of the following pairs of numbers:

        a) 36, 42   b) 218, 78   c) 142, 82   d) $12xy, 3x^2$

11. Look and read:

> In the number 1·23875, 2 is the first decimal place, 3 is the second decimal place and so on.
> 1·239 is the same number correct to three decimal places.
> 1·2 is the same number accurate to one decimal place.

Now ask and answer questions using the following table:

(**Note:** 0·333333...... may be expressed as 0·$\dot{3}$, or 0·3 recurring)

Tell me	$\sqrt{2}, \sqrt{5}, \frac{100}{9}, \frac{22}{7}$, etc.	correct	to	1 2 3 4 etc.	decimal place(s)
	the value of π, e, etc.	accurate			

Section 3   Reading

12. Copy this diagram, then label it as you read the passage:

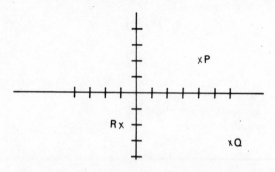

This diagram consists of two lines. The horizontal line is called the x-axis; the vertical line is called the y-axis. They intersect at the origin

18

O. Together, these two axes are called the co-ordinate axes. The axes separate the diagram into four quadrants. The top right-hand quadrant is known as the first quadrant, the top left the second, the bottom left the third, the bottom right the fourth. Starting from the origin, label the points to the right on the x-axis 1, 2, 3, 4, etc.; and to the left − 1, − 2, − 3, etc. On the y-axis, starting from the origin, label the points above the origin 1, 2, 3, 4, etc. and the points below the origin − 1, − 2, − 3, etc. Now any point in the plane may be represented in relation to the two axes by two numbers. For example, the point P is represented by two numbers, (4, 2). These numbers are called the co-ordinates of the point P. The x-co-ordinate, 4, is called the abscissa of P. The y-co-ordinate, 2, is called the ordinate of P. The system is known as the Cartesian co-ordinate system.

13. **Say whether the following statements are true or false. Correct the false statements.**

    a) The abscissa of the point Q is 3.
    b) The co-ordinates of the point R are (− 1, 2).
    c) The co-ordinates of the point P are (2, 4).
    d) The two axes divide the plane into four quadrants.

Section 4   Listening

The number system

14. **Listen to the passage and write down these words in the order in which you hear them:**

contains	consists	make up
composed	numerator	whereas
complex	symbol	cube
	any	

15. **Complete the following diagram:**

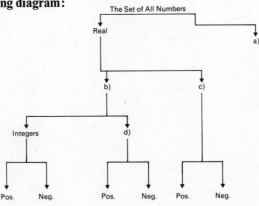

19

16. Say whether the following statements are true or false. Correct the false statements.

a) If m and n are integers, $\frac{m}{n}$ is a rational number.

b) The set of irrational numbers includes negative integers.

c) Neither irrational numbers nor complex numbers may be represented as points on a number line.

d) The symbol i represents a complex number.

e) A complex number consists of at least two parts.

17. Listen and describe the numbers which you will hear:

Example: 2·5: positive rational non-integer.

18. PUZZLE:

The display on an electronic watch consists of four digits, two for the hours and two for the minutes. At one minute past ten, the digits are the same from left to right as they are from right to left. How many times does this happen in 24 hours?

Unit A   Revision

1. Look and read:

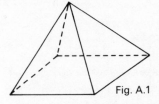

Fig. A.1

Figure A.1 shows a square pyramid.
The figure *is made up of* five faces.
The *bottom face* is a square.
Each lateral face is *a triangle with two sides equal.*
The point where the lateral sides meet *is called* the apex.
The distance from the apex to the centre of the square is 6 cm.
Each angle of the *bottom face* is *90°*.
The centre of the *bottom face* is *directly* below the apex.

Now replace the words in italics by words you already know.

2. Look at these:

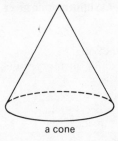

a cone

a frustum of a cone

Now write five sentences describing Figure A.2 using the words given:

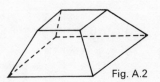

Fig. A.2

a) This is ......
b) ...... made up of ......
c) ...... lateral faces ...... shaped ......
d) ...... squares
e) ...... parallel.

3. Read this:

So far in this book we have looked at *properties*, *shapes*, *structure* and *location*. Look at each sentence in the following description and decide which of these four things it deals with.

a) The slide rule is rectangular in shape. b) It is usually made of wood

21

with a thin coating of plastic. c) It consists of three main parts: the body or stock, the slide, and the sliding marker or cursor. d) Both the slide and the cursor are moveable. e) The cursor is transparent. f) The slide rule has at least four basic scales: A, B, C and D. g) The B and C scales are located on the slide, whereas the A and D scales are situated on the stock. h) The C and D scales are identical.

4. Label this diagram:

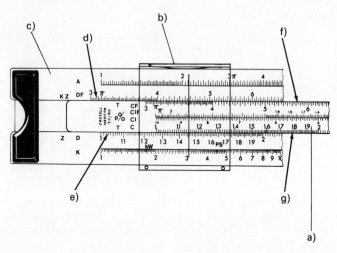

5. Now write a paragraph about computers, using the information given:

computers – hardware and software
software – programs; hardware – all functional parts
functional parts: input units, output units, memory units, control units, arithmetic units, auxiliary stores
memory, control, arithmetic – in one cabinet
memory, control, arithmetic = central processing unit
input, output, auxiliary = peripherals, outside CPU

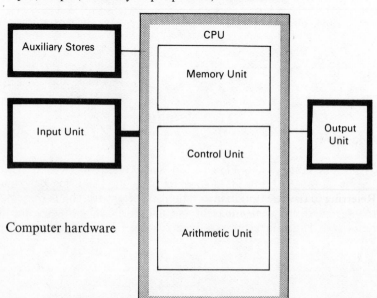

Computer hardware

Unit 4 Measurement 1

Section 1 Presentation

1. Read this:

Measurement of solid figures

Solid figures have three dimensions i.e. they are three-dimensional.
Rectangular prisms are solid figures. Figure 4.1 shows a right-rectangular prism i.e. a rectangular prism in which the lateral faces are perpendicular to the bases.

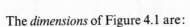

Fig. 4.1

The *dimensions* of Figure 4.1 are:

It has a height of 6 cm   It has a volume of 450 cm³
It has a width of 3 cm   It has a surface area of 486 cm²
It has a length of 25 cm   It has a cross-sectional area of 18 cm²

The *formula* for finding the volume is length × height × width. The formulae for finding the cross-sectional area and the surface area are, respectively, height × width and 2(hw + hl + wl).

Complete this description of the *structure* of a rectangular prism.

A rectangular prism consists of six ___a)___. The faces are divided into four _b)_ and two ___c)___. Each ___d)___ is ___e)___ in shape. ___f)___ faces are parallel.

2. Look at these:

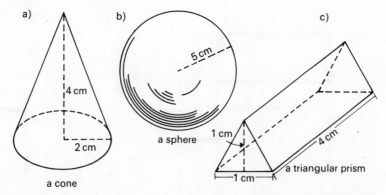

a) a cone

b) a sphere

c) a triangular prism

Referring to the description in exercise 1 of a right-rectangular prism, describe the *structure* and *dimensions* of the above solids, and give the formulae necessary for their measurement.

23

3. Read this short paragraph and draw the tube:

A cylindrical metal tube has a length of 12 cm and a cross-sectional diameter of 5 cm. The metal has a thickness of $\frac{1}{2}$ cm. Calculate the interior volume.

Now write similar paragraphs about the following and make the calculations:

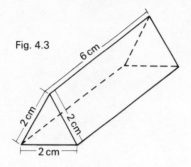

Fig. 4.3

6 cm

2 cm

2 cm

2 cm

a) A right _____ prism has an _____ triangle as its cross-section. Each _____ of the triangle _____ a length of 2 cm. The prism .... 6 cm. Draw the diagram and _____ the volume and surface _____.

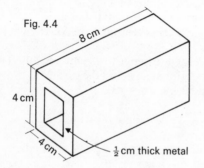

Fig. 4.4

8 cm

4 cm

4 cm

$\frac{1}{2}$ cm thick metal

b) i) interior volume?
    ii) surface area?

4. Look and read:

All faces of a regular polyhedron are congruent with each other. Only 5 kinds of regular polyhedrons exist.

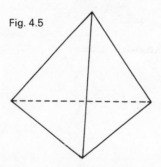

Fig. 4.5

A tetrahedron has 4 faces, 4 vertices and six *edges*. Each face is an equilateral triangle.

Now write similar sentences about the other regular polyhedrons.

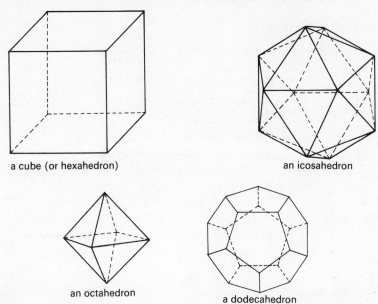

a cube (or hexahedron)

an icosahedron

an octahedron

a dodecahedron

5. Complete this table:

	Edges	Faces	Vertices
Tetrahedron	6	4	4
Cube			
Octahedron			
Dodecahedron			
Icosahedron			
Square pyramid			
Truncated square pyramid			
Pentagonal prism			

The relationship between edges, faces and vertices is a constant. Give this constant in a formula. It is known as Euler's formula.

Section 2   Development

6. Look at this table:

QUESTION	ANSWER
How high is that building?	It is 200 m high.
What is the height of that building?	The height of that building is 200 m. OR It has a height of 200 m.

Use the table to ask and answer questions about the following:
   (**Note:** The first pair is not always possible, and should not be used in (f) and (g)).

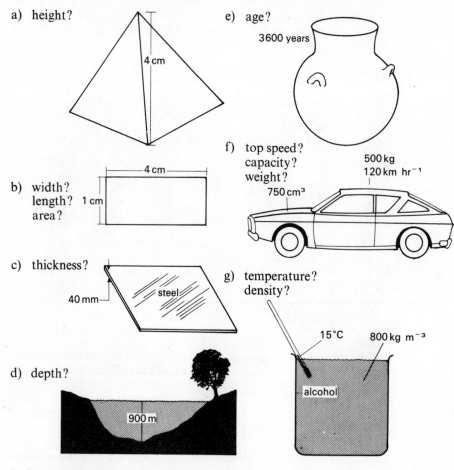

a) height?

4 cm

b) width?
    length? 1 cm
    area?

4 cm

c) thickness?

40 mm   steel

d) depth?

900 m

e) age?

3600 years

f) top speed?
    capacity?
    weight?
    750 cm³

500 kg
120 km hr⁻¹

g) temperature?
    density?

15°C   800 kg m⁻³

alcohol

7. Look and read:

   • Temperature is measured in degrees Celsius.
   • The degree Celsius is a unit of temperature.

26

Now make similar sentences beginning with the following words:

a) The joule   f) Velocity
b) The newton   g) Loudness
c) Time   h) Electric current
d) Mass   i) The litre
e) The watt   j) Angles

Section 3   Reading

8.   **Read this:**

Different kinds of average

The word 'average' is used frequently; for example, we talk about 'the average mark' in a test, the 'average rainfall' for a particular geographical location, or the 'average distance' of the Earth from the Sun' There are, however, many different kinds of average; here are some examples.

1   *The arithmetical mean*

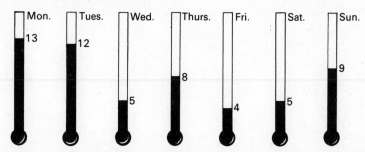

In a certain town, the temperature is recorded at mid-day each day in a particular week. The results are shown above. The 'average' is

$$\frac{13+12+5+8+4+5+9}{7} = 8,$$ and this is called the arithmetical mean.

2   *The midrange*

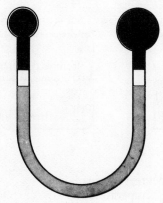

A maximum–minimum thermometer shows that the maximum temperature yesterday was $18°$ C and the minimum $2°$ C, i.e. the *range* is from $18°$ to $2°$ C, and $18°$ C and $2°$ C are the *extremes* of the range. The 'average'

$$\frac{18+2}{2} = 10° C$$ is called the mean of

the extremes or the midrange.

3 *The median*

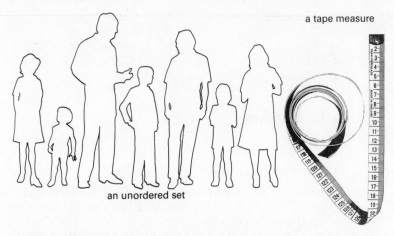

a tape measure

an unordered set

With a tape measure, we can find the arithmetical mean of the heights of the people shown above. Without a tape measure, we can use a different kind of average. First, we arrange the people in ascending or descending order of height. We then select the middle member of this ordered set. This kind of average is known as the median.

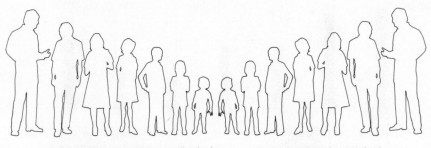

an ordered set (descending)   an ordered set (ascending)

4 *The mode*

Of the forty families who live in our street, 36 own a television set, 3 own 2 television sets, and one doesn't have a television set. If we take the arithmetical mean, we find that the 'average' family owns 1·05 television sets. A more useful average in this case is the mode i.e. the most common number, in this case one.

28

5   *The geometric mean*

1950   **?**   **1970**

1960

A man bought ten rabbits in 1950. In 1970, he had 200 rabbits. How many did he have in 1960? The answer is not the arithmetical mean of 10 and 200 i.e. 105, because population does not grow arithmetically (e.g. 2, 4, 6, 8, ...) but geometrically (e.g. 2, 4, 8, 16, 32, ...). The geometric mean of 10 and 200 is $\sqrt{10 \times 200} \simeq 45$.

9. Solve these problems:

a) Find the median, the mode, the arithmetical mean and the midrange of the following set of values.
   11   14   16   4   9   9   14

b) Find the geometric mean of the following pairs:
   3 and 27;   10 and 1000;   2 and 8192

10. Using words from the list, complete the definitions of the different kinds of average:

central	minimum	square root
divided	number	sum
divided	occurs	sum
maximum	ordered	value
	product	

a) The arithmetical mean of a set is the _____ of the members of the set _____ by the _____ of members.
b) The midrange of a set is the _____ of the _____ and _____ values of a set _____ by two.
c) The mode of a set is the _____ which _____ most frequently.
d) The median of a set is the _____ value of an _____ set.
e) The geometric mean of two numbers is the _____ of the _____ of the numbers.

11. Read and answer:

The following sets of numbers are arithmetical or geometrical *progressions*. Decide which is which and say why.

Example: $(6, 5, 4, 3, 2, 1)$ is an arithmetical progression because each element is formed by subtracting 1 from the preceding element.

a) $1, 2, 3, 4, 5, 6$

b) $2, 4, 8, 16, 32$

c) $\frac{1}{2}, \frac{1}{4}, \frac{1}{8}, \frac{1}{16}, \frac{1}{32}$

d) $x, (x+a), (x+2a), (x+3a)$

e) $11, 16, 21, 26, 31$

f) $x, \dfrac{x}{a}, \dfrac{x}{a^2}, \dfrac{x}{a^3}, \dfrac{x}{a^4}$

12. **Solve these problems:**

a) An arithmetical progression begins $\frac{1}{2}, \frac{1}{3}, \ldots\ldots$ What are the next two terms?

b) A geometrical progression begins $\frac{1}{2}, \frac{1}{3}, \ldots\ldots$ What are the next two terms?

Section 4 Listening

Vectors and scalars

13. **Listen to the passage and write down the word or phrase in each of the following pairs which occurs in the passage:**

the second quantity/a second quantity
both/boat
express/expressed
cylindrical/cylinder
40/30
both these/both of these
location/locating
consist/consists

14. **Divide the following quantities into vector or scalar quantities:**

speed, mass, displacement, weight, force, acceleration, velocity, distance, volume, temperature, momentum, power

15. **Say whether the following statements are true or false. Correct the false statements.**

a) The mass of an object is the same as the weight of the object.

b) An ordinary number is a vector quantity.

c) A vector quantity consists of two parts.

d) A position in the Cartesian co-ordinate system may be expressed by a vector.

16. PUZZLE

Comment on this statement:
A train travels 60 km at 30 km h^{-1}, then 60 km at 60 km h^{-1}. Its average speed is therefore 45 km h^{-1}.

Unit 5   Process 1   Function and Ability

Section 1   Presentation

1. Look and read:

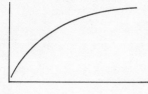

- A graph is a device for representing the relation between two variables diagrammatically.

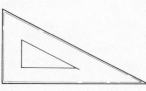

- A setsquare is an instrument for drawing or testing right angles.

Now make similar statements about the following:

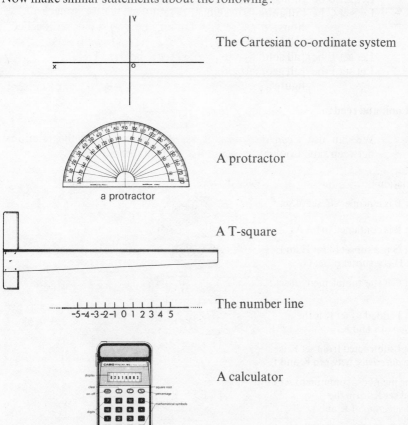

The Cartesian co-ordinate system

A protractor

a protractor

A T-square

The number line

A calculator

2. Look at these examples:

> A graph enables us to represent the relation between two variables diagrammatically.
> A setsquare enables us to draw or test a right angle.

Now make similar sentences using the diagrams in exercise 1.

Section 2   Development

3. Read this:

Relations between sets

> A set is any collection of things which we want to consider together. We use braces, { }, when we want to describe, or make a list of, the elements of a set, and we use capital letters to denote sets. For example:
>
> Let set A be {all animals}
> Let set B be {lions}
> Let set C be {all animals except lions}
> Let set D be {all human beings}
> Let set E be {all solid figures}
> Let set F be {all geometrical figures}
>
> Let set G be {positive integers < 10}
> Let set H be {1, 2, 3, 4, 5, 6, 7, 8, 9}
> Let set I be {1, 2, 3}
> Let set J be {∅} (This set is called the *empty* set)
> Let set K be {3, 4, 5}

4. Look and read:

> We can use *Venn diagrams* or *set notation* to show the relations between sets. Look at the examples in this table:

Relation	Venn diagram	Set notation
Set B is a proper subset of set A. Set B is contained in set A.		$B \subset A$
Set G is a subset of set H and set H is a subset of set G.		$G \subseteq H$ $H \subseteq G$
Set C is the complement of set B		$C = B'$
Set I added to set K is the *union* of I and K.		$I \cup K = \{1, 2, 3, 4, 5\}$
Set I subtracted from set K is the *difference* between K and I		$K - I = \{4, 5\}$
The members common to set I and set K form the *intersection* of K and I.		$K \cap I = \{3\}$
Sets B and D are *disjoint*.		$B \cap D = \{\emptyset\}$

Now make a similar table to express the following relations:

- a) A and D
- b) E and F
- c) H − I
- d) H ∪ I
- e) H ∩ I
- f) All triangles are plane figures.
- g) No triangles are solid figures.
- h) The set {figures with straight sides} and the set {figures with curved sides} have some members in common.

5. Using the following table, write definitions of the above relations between sets:

The union of two sets		a set which contains all the members which do not belong to A.
The intersection of two sets		a third set which contains all the members of both sets.
The difference between two sets	is	a set in which every member of A is also in B, but there is at least one member in B but not in A.
The complement of a set A		a third set which contains all the members of one set which are not common to both sets.
The subset A of a set B		a third set which contains all the members common to both sets.
The proper subset A of a set B		a set in which every member of A is also in B.

6. Read this:

The sign ∪ is used to symbolise union. A ∪ B is read as 'A union B'.

Now write similar sentences about the following signs:

- a) ∩
- b) −
- c) ′
- d) ⊂
- e) ⊃
- f) >
- g) ≤
- h) ∈

Section 3 Reading

7. Look and read:

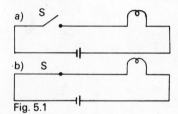

a)

b) S

Fig. 5.1

Figure 5.1 shows a simple electric circuit with a power source, a switch and a light. In (a), the switch is open, so the light is off. In (b), the switch is closed, so the light is on.

Now describe the following circuits in the same way:

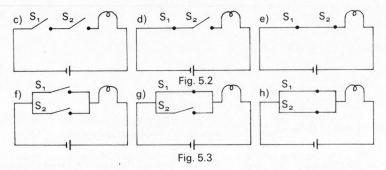

Fig. 5.2

Fig. 5.3

8. Read this:

The notation used to express relations in set theory is similar to the notation used to express relations in symbolic logic and circuit theory. For example, in Figure 5.1 we can use the letter p to denote the statement 'S is open', so p describes circuit (a), when the light is off. The notation p' i.e. the complement or the negation of p, means 'S is not open', so p' describes circuit (b), when the light is on.

In the circuit shown in Figure 5.2, there are two switches, S_1 and S_2. Let p denote the statement 'S_1 is open' and let q denote the statement 'S_2 is open'. The circuit as shown in Figure 5.2(c) can now be described as 'p and q' or, using logic notation, as $p \cap q$ or $p \wedge q$. Closing both switches allows current to flow (Figure 5.2(e)). The expression $p' \wedge q'$ accurately describes this circuit. Current cannot flow if either (i) one switch is open, or (ii) both switches are open, i.e. there are three possibilities: $p \wedge q$, $p \wedge q'$ and $p' \wedge q$. These possibilities are described by the expression p or q. This is expressed in logic notation as $p \cup q$ or $p \vee q$.

Look at this example:

- The circuit in Figure 5.1(a) can be described as p.

Now write similar sentences describing the circuits in the following figures:

5.1, 5.2, 5.3 and 5.4. In Figure 5.3, let $p = S_1$ is open, $q = S_2$ is open. In Figure 5.4, let $p = S_1$ is in position 1, $q = S_2$ is in position 1.

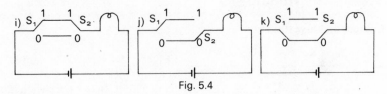

Fig. 5.4

9. Complete the following table using logic notation:

Circuit	Current flows	Current does not flow
5.1	p'	p
5.2		
5.3		
5.4		

10. Look at these examples:

- In Figure 5.4, putting both switches in the same position allows current to flow.
- In Figure 5.4, putting the switches in opposite positions prevents current from flowing.

Now write similar sentences about the other circuits.

Section 4   Listening

Sets of numbers

11. Listen to the passage and write down these words in the order in which you hear them:

common	frequently	referred to
differs	includes	usually
enables	oblique	whereas

12. Listen to the passage again and write down the symbols for the following sets as you hear them:

a) the empty set
b) the universal set
c) all integers
d) all rational numbers
e) all natural numbers
f) all real numbers

13. Draw a Venn diagram representing the relations between the four sets (c), (d), (e) and (f) in exercise 12.

Using your diagram, say whether the following statements are true or false. Correct the false statements.

a) $N \subset R$
b) $N \subseteq R$
c) $N \subseteq Z^+$
d) $N \subset Z^+$
e) $Q \supset R$
f) $R - Q =$ the set of all imaginary numbers
g) There are no disjoint sets in the diagram

14. PUZZLE

Consider these statements:

> A dog is an animal
> A cat is an animal
> Therefore a dog is a cat

Which two mathematical symbols can be used for the different meanings of 'is' to enable us to see the *flaw* in the above argument?

Unit 6　Process 2　Actions in Sequence

Section 1　Presentation

1. Look and read:

$1+2+3+4+5+6+\ldots\ldots$

- As successive values are added to this series, so the sum gets larger and larger.

$\frac{1}{2}+\frac{1}{4}+\frac{1}{8}+\frac{1}{16}+\frac{1}{32}+\ldots\ldots$

- As successive values are added to this series, so the sum approaches 1.

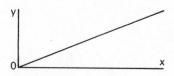

- As x becomes larger, so y becomes larger.

Complete the following sentences in the same way:

$2+4+8+16+32+\ldots\ldots$　　a)　As successive values ......
$1+\frac{1}{2}+\frac{1}{4}+\frac{1}{8}+\frac{1}{16}+\ldots\ldots$　　b)　As successive values ......
$1!+2!+3!+4!+5!+\ldots\ldots$　　c)　As successive values ......

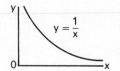

d)　As x becomes larger, ...
e)　As x becomes smaller, ...

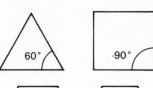

f)　As the number of sides of regular polygons is increased, so the angles ......

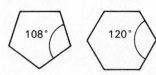

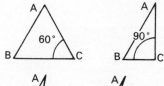

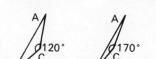

g)　As angle C approaches 180°, so angles A and B ......

$$\sqrt{2} = 1\cdot414 \quad \sqrt[3]{2} = 1\cdot2599$$
$$\sqrt[6]{2} = 1\cdot1225 \quad \sqrt[8]{2} = 1\cdot0905$$
$$\sqrt[14]{2} = 1\cdot0508$$

h) As successive roots of 2 are taken, ...

Section 2 Development

2. Read this:

- In the set of real numbers, how large is the highest number?
  However large a number is, there is always a higher number.
- In the set of numbers <1, what number is the highest member of the set?
  Whatever number we choose, there is always a higher number in the set.
- How many points are there on a line?
  However many points we choose, there are always more points.

Now make correct statements from the table:

In the set of real numbers		large we make one angle	there is always a smaller value.
On a line		distance we take between two points	the sum does not reach one.
In the set $x>0$	however	small a number is	there is always a shorter distance.
In the series $\frac{1}{2}+\frac{1}{4}+\frac{1}{8}+\dots$	whatever	many values we add	it cannot be more than $180°$.
In the series $\sqrt{2}, \sqrt[3]{2}, \sqrt[4]{2}, \sqrt[5]{2}, \dots$		root of 2 is taken	there is always a smaller number.
In a triangle		value of x we take	its value is always greater than one.

3. Look and read:

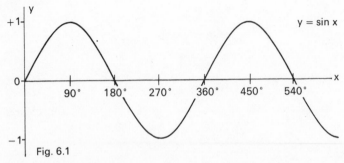

Fig. 6.1

Figure 6.1 is a graph of the function $y = \sin x$. As x goes from $0°$ to $90°$, $\sin x$ increases from 0 to 1. As x goes from $90°$ to $270°$, $\sin x$

decreases from 1 to -1, crossing the x-axis at 180°.
As x goes from 270° to 360°, sin x increases from -1 to 0.
When x reaches 360° the graph repeats itself. The sine function is a periodic function, with a period of 360°, i.e. the graph repeats itself every 360°.

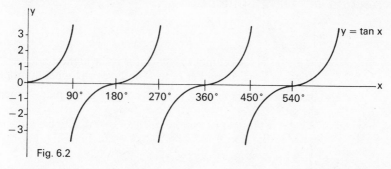

Fig. 6.2

Figure 6.2 is a graph of the function $y = \tan x$. As x approaches 90°, tan x tends to infinity. After 90°, tan x reappears on the negative side. As x goes from 90° to 180°, tan x increases to 0. As x approaches 270°, tan x again tends to infinity, reappearing again after 270° on the negative side. The tangent function is a periodic function, with a period of 180°, i.e. the graph repeats itself every 180°.

Now describe the following trigonometrical functions:

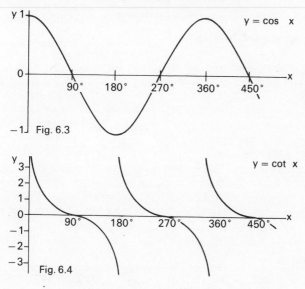

Fig. 6.3

Fig. 6.4

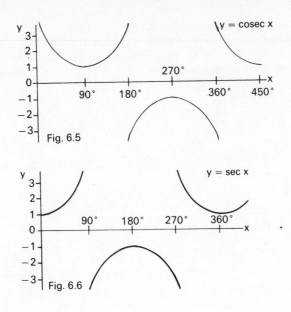

Fig. 6.5

Fig. 6.6

Section 3　Reading

4. Read this:

Convergence and divergence

An expression of the form $a_1 + a_2 + a_3 + a_4 + \ldots + a_n + \ldots$ is called an *infinite series*, or simply a *series*. $a_1, a_2, a_3, a_n, \ldots$ are called, respectively, the first, second, third, nth, etc. *terms*. Each term in a series can be calculated from the preceding term by using a given rule. For example, in the series $1 + 2 + 3 + 4 + \ldots$, each term is found by adding one to the preceding term.

Although the number of terms in a series is infinite, the sum of the terms may have a finite *limit*. For example, the sum of the series $1 + \frac{1}{2} + \frac{1}{4} + \frac{1}{8} + \ldots$, where each term is found by multiplying the preceding term by $\frac{1}{2}$, gets nearer and nearer to 2 but never reaches it. 2 is consequently said to be the limit of the series, and the series is said to be *convergent*.

A series in which the sum does not tend to a finite limit is said to be *divergent*, as in the series $1 + 2 + 3 + 4 + \ldots$

In all convergent series, the terms get closer and closer to zero, but not all series in which the terms get closer and closer to zero are convergent. For example the terms of the series $1 + \frac{1}{2} + \frac{1}{3} + \frac{1}{4} + \frac{1}{5} + \ldots$ get closer and closer to zero, but the sum increases without bound. This can be seen if we re-write the series as $1 + \frac{1}{2} + (\frac{1}{3} + \frac{1}{4}) + (\frac{1}{5} + \frac{1}{6} + \frac{1}{7} + \frac{1}{8}) + (\frac{1}{9} + \frac{1}{10} + \ldots + \frac{1}{16}) + (\frac{1}{17} + \frac{1}{18} + \ldots \frac{1}{32}) + \ldots$ Each sum in the brackets is greater than $\frac{1}{2}$, so the sum of the series is always greater than $1 + \frac{1}{2} + \frac{1}{2} + \frac{1}{2} + \ldots$, and the series is divergent.

Say whether the following statements are true or false. Correct the false statements.

    a) Any term in a series is always positive.
    b) All series are either convergent or divergent.
    c) A convergent series increases without bound.
    d) The sum of the series $1 + \frac{1}{2} + \frac{1}{4} + \ldots$ tends to zero.
    e) Whatever term in a series we choose, it is always possible to add more terms.
    f) In convergent series, the terms get smaller.

Section 4 Listening

Stationary points

5. **Listen to the passage and write down the word in each of the following pairs which occurs in the passage:**

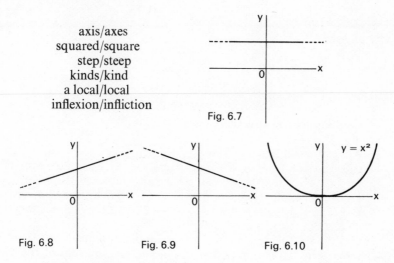

axis/axes
squared/square
step/steep
kinds/kind
a local/local
inflexion/infliction

Fig. 6.7

Fig. 6.8

Fig. 6.9

Fig. 6.10

6. **Complete this table:**

	Gradients	
	Before	After
Maximum	+	−
Minimum		
Point of inflexion Either		
Or		

7. **Say whether the following statements are true or false. Correct the false statements.**

    a) The sine function has a turning point every 90°.
    b) The cosecant function has a point of inflexion at 270°.
    c) In the function $y = x^2$, the gradients at $x = 1$ and $x = -1$ are equal.
    d) The tangent to a curve at a stationary point can have either a positive or a negative gradient.

8. PUZZLE

A man has three sons. When left alone with one of his brothers, the oldest son always starts a fight. The man needs to cross a river using a boat which can carry only two people. He also wants to avoid a fight if possible. Can he do it?

Unit B Revision

1. Draw a Venn diagram to illustrate the relationships between the following sets:

  Set A: {all plane figures}
  Set B: {circles, triangles, squares}
  Set C: {all figures with curves or curved surfaces}
  Set D: {all solid figures}
  Set E: {cylinders, octahedra, pyramids}

Now answer these questions:

  a) What does the union of B and C consist of?
  b) What are the elements of the intersection of C and E?
  c) What set is formed by the intersection of A and D?
  d) Name some members of the intersection of C and (A − B)?
  e) Name a pair of disjoint sets.
  f) Put your results for (a) to (e) into mathematical form.

2. Look and read:

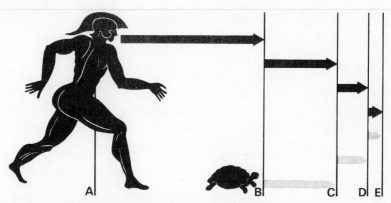

Zeno, a Greek philosopher who lived 2 500 years ago, posed the following problem. Achilles, a fast runner, has a race against a tortoise. Before the race, Achilles is at point A and the tortoise at point B. After ten seconds, Achilles reaches point B, but the tortoise has advanced to point C. When Achilles reaches point C, the tortoise has advanced to point D, and so on. In this way, Achilles can never reach the tortoise.

Now complete these sentences using the following words:

  a) Before starting the race ......

b) As Achilles runs from A to B, ...
c) As the race progresses, the distances between successive points ......
d) According to Zeno, however fast Achilles runs, ...

What is the flaw in Zeno's argument?

3. Read this:

What happens when two bodies collide? We can divide the event into three periods: before, during and after impact.
For example, sphere A and sphere B have the same size, but different masses, denoted by m_a and m_b respectively.

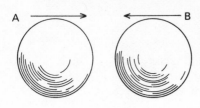

They move towards each other at different speeds. Their velocities are in opposite directions. Therefore, using the letter u to denote initial velocity, we can say that sphere A has a velocity of u_a in the positive direction, while sphere B has a velocity of u_b in the negative direction.
On impact, sphere B exerts a force on sphere A and sphere A exerts an equal and opposite force on sphere B. The product of this force and the duration of the impact (force × time) is called the impulse. The time of impact is very small, but whatever the duration of the time of impact, the impulses are equal and opposite, since the forces are equal and opposite.
After impact, the spheres have new velocities, v_a and v_b, respectively.

The momentum of a body is given by the product of its mass and its velocity, and impulse can be measured by subtracting the initial momentum from the final momentum, i.e. $ft = mv - mu$. The impulses on spheres A and B are equal but opposite, and so $mv - mu$ (for A) $= -(mv - mu)$ (for B),
i.e. $m_a v_a - m_a u_a = m_b u_b - m_b v_b$.
Therefore $m_a v_a + m_b v_b = m_b u_b + m_a u_a$
i.e. no momentum is lost as a result of the collision.

Answer these questions:

a) What is the mass of sphere A denoted by?
b) Why is the velocity of sphere B negative?
c) What kinds of quantities are force, momentum, impulse and velocity?
d) What are the following measured in:

     i) Force?
    ii) Time?
  iii) Impulse?

4. Now write a short paragraph about the following:

During impact, two periods: a) spheres change shape
                                   b) spheres regain shape

Ability to regain shape = elasticity
  e.g.   hard metal – high elasticity
           soft metal – low elasticity

Unit 7   Measurement 2   Quantity

Section 1   Presentation

1. Look and read:

Exact calculations and approximations

Some square roots may be calculated exactly

$$e.g. \quad \sqrt{4} = 2$$
$$\sqrt{6\cdot25} = 2\cdot5$$
$$\sqrt{14\cdot44} = 3\cdot8$$

Other square roots may be calculated only approximately

$$e.g. \quad \sqrt{2} = 1\cdot414213\ldots\ldots$$
$$\sqrt{3} = 1\cdot7320508\ldots\ldots$$
$$\sqrt{5} = 2\cdot236068\ldots\ldots$$

These approximate square roots are called irrational numbers i.e. we can continue the numbers after the decimal point as long as we wish.

Look at the following and say whether they can be calculated *exactly* or *only approximately*:

a) $\sqrt{9}$ c) $\sqrt{12\cdot25}$ e) The area of a circle   g) $\sqrt{110\cdot25}$

b) $\sqrt{13}$ d) π f) Any irrational number   h) $\sqrt{23\cdot5}$

2. Read this:

Approximations to square roots

To find $\sqrt{7}$:
First, we guess a value for $\sqrt{7}$, say $2\frac{1}{2}$.
$7/2\frac{1}{2} = 2\cdot8$. Thus $2\frac{1}{2}$ is too small.
So we try a value half-way between $2\frac{1}{2}$ and $2\cdot8$ i.e. $2\cdot65$.
$7/2\cdot65 = 2\cdot64$. Thus $2\cdot65$ is slightly too large.
So we try $(2\cdot65 + 2\cdot64)/2 = 2\cdot645$.
$7/2\cdot645 = 2\cdot646$.
$\sqrt{7}$ may be calculated to an *arbitrary* degree of accuracy i.e. we can calculate· it to *any required* degree of accuracy, but $2\cdot645$ is a reasonably good approximation.

Now write similar paragraphs using the following examples:

a) $\sqrt{11}$; first guess $3\frac{1}{2}$ b) $\sqrt{34}$: first guess $5\frac{1}{2}$

3. Read this:
- 4 exceeds $\sqrt{7}$ by a considerable amount.
- 2·65 exceeds $\sqrt{7}$ by a very small amount.

Make similar statements about the following pairs:

a) $6; \sqrt{4}$ b) $2 \cdot 24; \sqrt{5}$ c) $22/7; \pi$ d) $3 \cdot 2; \sqrt{5}$

Section 2 Development

4. Look and read:

$0 < x < 1$	No integral value of x satisfies this inequality.
$x = \sqrt{-1}$	No real number satisfies this equation.
x is a prime number and x is even	Only one number satisfies these requirements.

Now write similar sentences about the following:

x is the square of an integer and the last digit of x is 3	a) ...... value ......
$x^2 + 5 = 0$	b) ...... real number ......
$0 < x < 2$	c) ...... integral value ......
$x^2 + 2x - 35 = 0$	d) ...... positive value ......
$x = \sqrt{2}$	e) ...... rational number ......
x is divisible by both 7 and 9 and $x < 100$	f) ...... value ......

5. Look and read:

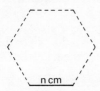
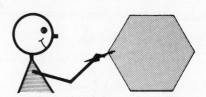

We are given the length of one side of a regular hexagon. This is sufficient for the area to be calculated.

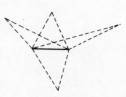

We are given the length of one side of a triangle. This is insufficient for the area to be calculated.

Now write sentences about the following in the same way:

Given	Required
a) one side of a square	area
b) one side of a rectangle	area
c) the altitude of a cone	volume
d) the area of one face of a regular dodecahedron	surface area
e) the length of the non-parallel sides of a trapezium	area
f) the surface area of a sphere	volume
g) the area of the lateral faces of a prism	volume
h) a chord of a circle	area

6. Look and read:

Congruence of triangles

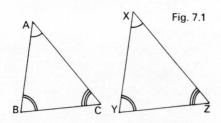

Fig. 7.1

- In Figure 7.1 $\widehat{A} = \widehat{X}$, $\widehat{B} = \widehat{Y}$, $\widehat{C} = \widehat{Z}$ (i.e. the angles are equal). This is a *necessary* condition for the two triangles to be congruent, but it is not a *sufficient* condition, i.e. the two triangles may be congruent, but we have insufficient information.

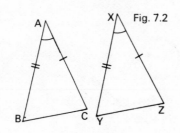

Fig. 7.2

- In Figure 7.2 $\widehat{A} = \widehat{X}$, AB = XY, AC = XZ (i.e. two sides and the included angle are equal). This is a sufficient condition for the triangles to be congruent, i.e. the triangles are congruent.

Now write about the following pairs of triangles in the same way:

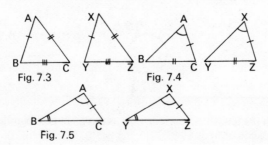

Fig. 7.3   Fig. 7.4

Fig. 7.5

48

Section 3   Reading

7. Read this:

The solution of triangles

A triangle has three sides and three angles. When three of these elements are known and at least one of the elements is a side, the other three elements can be calculated. Only one trigonometrical ratio, sine, is required in the calculation.

In a triangle ABC, we are given the lengths of AB and AC and the value of $A\widehat{B}C$.

We use this formula:

$$\text{Sin } A\widehat{C}B = \frac{AB \sin A\widehat{B}C}{AC}$$

The fraction $\dfrac{AB \sin A\widehat{B}C}{AC}$ may be of three kinds:

    i) an improper fraction i.e. greater than one;
    ii) a proper fraction i.e. less than one;
or iii) exactly equal to one.

In case (i), AC is smaller than AB sin $A\widehat{B}C$. This requires sin $A\widehat{C}B$ to be greater than one, which is impossible. Therefore no such triangle can exist.

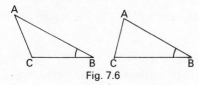

Fig. 7.6

In case (ii), AC is greater than AB sin $A\widehat{B}C$. Two values of $A\widehat{C}B$ may satisfy the equation (Figure 7.6). In this case, further information is required to solve the triangle exactly.

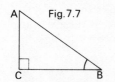

Fig. 7.7

In case (iii), AC is equal to AB sin $A\widehat{B}C$. Only one solution satisfies the equation: $A\widehat{C}B = 90°$ (Figure 7.7).

Say whether the following statements are true or false? Correct the false statements.

    a) The sine ratio is sufficient for triangle ABC (given AB, AC and $A\widehat{B}C$) to be solved.
    b) Any three elements of a triangle are sufficient for it to be solved.
    c) Case (i) would require $A\widehat{C}B$ to be greater than two right angles.
    d) Only one further element is required to solve the triangle in (ii).
    e) In (ii) we are given the further information that $A\widehat{C}B$ exceeds one right angle. This is sufficient for the triangle to be solved.

8. **Look at these examples:**

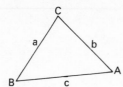

- $\widehat{A} > 180°$. No such triangle can exist.
- $a = b = c = 3\,cm$. Only one such triangle can exist.

Write similar sentences about the following cases:

a) $\dfrac{b \sin A}{a} > 1$

b) $\dfrac{b \sin A}{a} \geqslant 1$

c) $\dfrac{b \sin A}{a} = 1$

d) $\dfrac{b \sin A}{a} < 1$

Section 4 Listening

Approximate values

9. **Listen to the passage and write down in figures each number you hear.**

10. **Listen to the passage again and say whether the following statements are true or false. Correct the false statements.**

    a) 3·1416 is an approximate value of π.
    b) The difference between two approximate values is known as the absolute error.
    c) The absolute error is the same as the true error.
    d) The relative error is the true value divided by the absolute error.
    e) The percentage error is found by multiplying the absolute value by 100.
    f) The true value of 3·76, which is accurate to three significant figures, may be anywhere between 3·7 and 3·8.

11. **Look at this example:**

3·76 is an approximate value of 3·757 accurate to three significant figures.

Now make similar sentences about the following:

    a) 3·1416; π
    b) 0·108; 0·1077
    c) 3 500; 3 498

12. Solve these problems:

Find a) the absolute error, b) the relative error and c) the percentage error in exercise 11 c).

13. PUZZLE:

How many different digits are needed to give the value of:

a) $1/3$ b) $(1/3)^2$ c) $(1/3)^4$ to ten significant figures?

Unit 8   Process 3   Cause and Effect

Section 1   Presentation

1. Read this:

Quadratic equations

A quadratic expression is an expression which contains a number raised to the power of 2 (e.g. x^2). It cannot contain numbers raised to powers greater than 2 (e.g. x^3, x^4, etc.)

Say which of the following are quadratic expressions:

a) $x^2 + 3$
b) $x^2 + 3x + 7$
c) $3x^2 - 2$
d) $x^2 + x^4$
e) $x + 2y + z$
f) $x^3 + 2x - 16$

2. Read this:

A quadratic expression is generally given in the form $ax^2 + bx + c$, where x is the variable and a, b and c are constants. A quadratic equation is generally given in the form $ax^2 + bx + c = 0$.

Now change the following equations to the general form for quadratic equations and give the values of a, b and c. The first two are done for you:

	Given	General form	a	b	c
a)	$2x^2 - 3x = 2$	$2x^2 - 3x - 2 = 0$	2	-3	-2
b)	$3x^2 = -1$	$3x^2 + 0x + 1 = 0$	3	0	1
c)	$x^2 = 3x$				
d)	$5x - 3 = 4x^2$				
e)	$x^2 + x = 1$				
f)	$5x^2 + 7 = 20$				

3. Look at this:

A quadratic equation has two solutions, called roots. If the factors of a quadratic equation can be found easily, then we can find the roots by factorising.

Example: Factorisation of $x^2 + x - 12 = 0$ gives $(x - 3)(x + 4) = 0$.
The roots of the equation are therefore 3 and -4.

Now make similar sentences about the following:

a) $x^2 + 7x + 10 = 0$
b) $x^2 - 9x + 18 = 0$
c) $x^2 - 100 = 0$
d) $x^2 + 5x - 6 = 0$

4. Read this:

Factorisation of $x^2 + 12x + 36$ gives $(x + 6)^2$. Therefore the expression is known as a perfect square.

$x^2 + ax$ can be made into a perfect square by adding $\left(\dfrac{a}{2}\right)^2$.

For example, $x^2 + 20x$ can be made into a perfect square by adding 100.
$x^2 + 20x + 100$ factorises into $(x + 10)^2$.

Write similar sentences about the following expressions:

a) $x^2 - 12x$

b) $x^2 + 3x$

c) $x^2 + 7x$

(**Note:** This operation is known as *completing the square*).

Section 2   Development

5. Look and read:

If the factors of a quadratic equation cannot be found easily, then we can find the roots by using the formula

$$\frac{-b \pm \sqrt{(b^2 - 4ac)}}{2a}$$

The two roots are at $\dfrac{-b + \sqrt{(b^2 - 4ac)}}{2a}$ and $\dfrac{-b - \sqrt{(b^2 - 4ac)}}{2a}$

If $(b^2 - 4ac)$ is negative, then $\sqrt{(b^2 - 4ac)}$ is imaginary and no real roots satisfy the equation.

Complete these two sentences:
a) If $(b^2 - 4ac)$ is positive, ...
b) If $(b^2 - 4ac)$ is zero, ...

6. Look and read:

The two roots of a quadratic equation are denoted by α and β. We can show the six possible cases by drawing graphs.

53

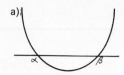

a)

- If $(b^2 - 4ac)$ is positive and a is positive, then there are two real roots at α and β and the quadratic expression is only negative for values of x between α and β.

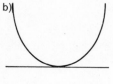

b)

- If $(b^2 - 4ac)$ is zero and a is positive, then the two real roots α and β *coincide*, and the quadratic expression is positive for all other values of x.

c)

- If $(b^2 - 4ac)$ is negative and a is positive, then there are no real roots, and the quadratic expression is always positive.

Now describe the other three cases:

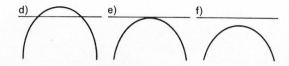

d)   e)   f)

7. Look at these examples:

$x^2 - 2x - 3 = 0$ i) Factorisation of the left-hand side gives $(x-3)(x+1) = 0$.

ii) Factorising the left-hand side gives $(x-3)(x+1) = 0$.

Now change the following examples to form (ii):

a) $x^2 - 2x - 3 = 0$ Addition of 4 to both sides gives a perfect square.

b) $\dfrac{25}{10}$ Reduction of the fraction gives $\dfrac{5}{2}$.

c) $9x = 18y$ Division of both sides by 9 gives $x = 2y$.

d) $x^2 + 10x + 32$ Subtraction of 7 from this expression gives a perfect square.

e) $x^2 - 5x + 6 = 0$ Solution of this equation gives roots at 2 and 3.

f) $a - \dfrac{a}{x^2} = 0$ Multiplication of both sides by x^2 gives $ax^2 - a = 0$.

8. Look at this example:

$x^2 - 10x - 200 = 0$.
Factorising, we obtain $(x+10)(x-20) = 0$

54

Use expressions from this list to complete the calculation below.

completing the square,
dividing ......,
factorising,
subtracting ......,
subtracting ......,
taking the square root ......,

Given $ax^2 + bx + c = 0$

...... we obtain $x^2 + \dfrac{b}{a}x + \dfrac{c}{a} = 0$.

...... we obtain $x^2 + \dfrac{b}{a}x = -\dfrac{c}{a}$

...... we obtain $x^2 + \dfrac{b}{a}x + \left(\dfrac{b}{2a}\right)^2 = \left(\dfrac{b}{2a}\right)^2 - \dfrac{c}{a}$

...... we obtain $\left(x + \dfrac{b}{2a}\right)^2 = \left(\dfrac{b}{2a}\right)^2 - \dfrac{c}{a}$

...... we obtain $x + \dfrac{b}{2a} = \dfrac{\pm\sqrt{(b^2 - 4ac)}}{2a}$

...... we obtain $x = \dfrac{-b \pm \sqrt{(b^2 - 4ac)}}{2a}$

This gives the formula for finding the roots of a quadratic equation.

Section 3   Reading

9. Look and read:

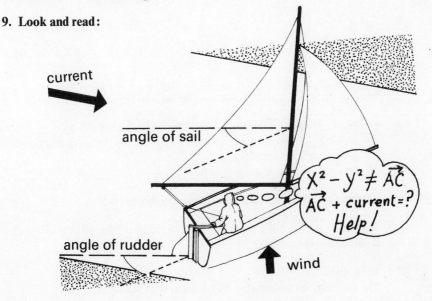

current

angle of sail

angle of rudder

wind

$X^2 - y^2 \neq \overrightarrow{AC}$
$\overrightarrow{AC} + current = ?$
Help!

Addition of vectors

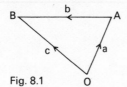

Fig. 8.1

Taking O as the origin, we can determine any position A by naming the length and direction of the line which joins O to A. Thus the vector a in Figure 8.1 is denoted by $\overrightarrow{OA}$ or $\overline{OA}$.

We may reach a second position, B, by displacement first from O to A and then from A to B. But direct displacement from O to B gives the vector $\overline{OB}$. By comparing these two results, we see that $\overline{OA} + \overline{AB} = \overline{OB}$ (1). If we complete the parallelogram OABC, then $\overline{CB} = \overline{OA}$, and $\overline{OC} = \overline{AB}$.

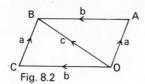

Fig. 8.2

Replacing $\overline{AB}$ by $\overline{OC}$ in equation (1) gives $\overline{OA} + \overline{OC} = \overline{OB}$ (2). This gives the parallelogram law for the addition of two vectors. $\overline{OB}$ is known as the resultant of $\overline{OC}$ and $\overline{OA}$.

We can repeat this addition as often as we like so that we can find the resultant of any number of vectors.

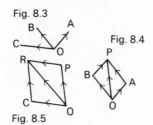

Fig. 8.3

Fig. 8.4

Fig. 8.5

If we want to find the resultant of the three vectors $\overline{OA}$, $\overline{OB}$ and $\overline{OC}$ (Fig. 8.3), then we first add $\overline{OA}$ and $\overline{OB}$ giving the resultant $\overline{OP}$ (Fig. 8.4). Adding the resultant $\overline{OP}$ and the third vector $\overline{OC}$ (Fig. 8.5), we have $\overline{OA} + \overline{OB} + \overline{OC} = \overline{OR}$.

10. **Say whether the following statements are true or false. Correct the false statements.**
    a) In Figure 8.1, subtracting AB from OB gives AO.
    b) Equation (1) means that two sides of a triangle can be equal to the third side.
    c) By addition of $\overline{BC}$ and $\overline{OA}$ in Figure 8.2, we obtain $\overline{AC}$.
    d) Taking the vectors $\overline{OA}$, $\overline{OB}$ and $\overline{OC}$ (Fig. 8.3) in a different order gives a different resultant.
    e) In Figure 8.2, $a + b - c =$ zero.

11. **Look at these sentences and put them in the correct order:**

    They are taken from a paragraph of instructions for constructing the resultant of OA and OB in Figure 8.3.

    a) Join OP.
    b) Then, with radius OB, draw an arc with centre A.

c) $\overline{OP}$ is the desired resultant.

d) The second arc cuts the first at P.

e) With radius OA, draw an arc with centre B.

12. Look at this example:

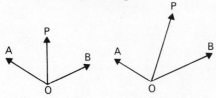

The effect of making $\overline{OB}$ longer is to make $\overline{OP}$ longer.

Now write similar sentences about the following:

a) The effect on parallelogram OAPB of making $\overline{OA}$ have the same magnitude as $\overline{OB}$.

b) The effect on angles AOP and BOP of making $\overline{OA}$ have the same magnitude as $\overline{OB}$.

c) The effect on parallelogram OAPB of making $\overline{OA} = \overline{OB}$.

d) The effect on the resultant of adding $\overline{OA}$ and $\overline{BO}$.

Section 4 Listening

Multiplication of vectors by scalars

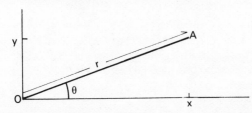

Fig. 8.6

13. Listen to the passage and write down these words in the order in which you hear them:

either	opposite	polar
alter	unchanged	process
negative	multiply	way

14. Listen again and complete this exercise:

a) Sketch a Cartesian co-ordinate system showing the effect of multiplying $\overline{OA}$ in Figure 8.6 by 3.

b) What are the polar co-ordinates of the resultant vector?

c) What are the rectangular co-ordinates of the resultant vector?

d) If $\overline{OA}$ is multiplied by a negative scalar quantity, in which quadrant is the resultant vector?

e) What are the rectangular co-ordinates of any vector (x,y) multiplied by $-a$?

15. Look at this example:

Adding two vectors with different directions produces a change in direction.

Now make similar sentences about the following:

a) Multiplying a vector by a scalar ...... (magnitude)
b) Multiplying a vector by a scalar ...... (direction)
c) Multiplying a vector by a negative scalar ...... (angle)
d) Multiplying a vector by a negative scalar ...... (direction)

16. PUZZLE

Comment on the following:

Given: $a > b$

c is the arithmetic mean of a and b.

Therefore $c = \dfrac{a+b}{2}$

or $a + b = 2c$

Multiplying both sides by $a - b$, we obtain

$$(a+b)(a-b) = 2ac - 2bc$$

i.e. $a^2 - b^2 = 2ac - 2bc$

Adding b^2 to both sides gives $a^2 = 2ac - 2bc + b^2$
Adding c^2 to both sides gives $a^2 + c^2 = 2ac - 2bc + b^2 + c^2$
Subtracting 2ac from both sides gives $a^2 - 2ac + c^2 = b^2 - 2bc + c^2$
Factorisation of both sides produces $(a-c)^2 = (b-c)^2$
Taking the square root, $a - c = b - c$
Adding c to both sides $a = b$

Unit 9   Measurement 3
Ratio and Proportion

Section 1   Presentation

1. Look at these examples:

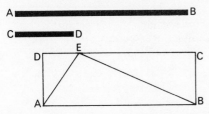

AB is approximately three times as long as CD.

The area of rectangle ABCD is exactly twice as big as that of $\triangle$ABE.

Now make similar sentences about the following:

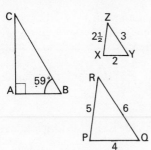

a) $\hat{ABC}$ ...... $\hat{ACB}$.

b) PR ...... XZ.

c) $\triangle$PQR ...... $\triangle$XYZ.

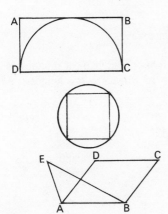

d) ABCD ...... the inscribed semicircle.

e) ...... any circle ...... the inscribed square.

f) The circumference ...... the diameter.

g) ...... rhombus ABCD ...... ABE.

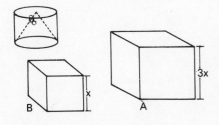

h) ...... cylinder ...... cone with the same base and height.

i) ...... cube A ...... cube B.

59

2. Look at these examples:

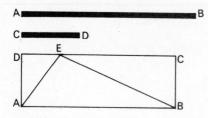

The lengths of AB and CD are in the ratio of approximately 3:1.

The areas of ABCD and ABE are in the ratio of exactly 2:1.

Now make similar sentences about the examples in exercise 1.

Section 2   Development

3. Look and read:

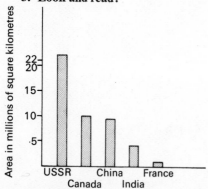

This bar graph shows the relative sizes of some countries in millions of square kilometres.

- The USSR is far larger than India.
- India is considerably larger than France.
- Canada is slightly larger than China.

Now compare the other countries in the same way.

4. Look at the table below showing the heights of the highest mountains in the different continents. Draw a bar graph to illustrate the heights and then compare the heights of different mountains:

Continent	Highest Mountain	Height (in metres)
Africa	Kilimanjaro	5 963
Antarctica	Vinson Massif	5 139
Asia	Everest	8 880
Australasia	Wilhelm	4 693
Europe	Elbrus	5 633
North America	McKinley	6 187
South America	Aconcagua	6 959

5. Look and read:

Variation

$y \propto x$ The ratio between y and x is a constant.
y is *directly proportional* to x.
We say that y *varies directly* as x.

$y \propto \dfrac{1}{x}$ y is directly proportional to the reciprocal of x.
We say that y is *inversely proportional* to x and that y *varies inversely* as x.

$y \propto xz$ y is directly proportional to the product of x and z.
We say that y is *jointly proportional* to x and z, and that y *varies jointly as* x and z.

Describe the relationship between the following quantities, where k is a constant:

   a) $y = \dfrac{k}{x}$
   b) $y = kxz$
   c) $y:x = k$
   d) $y = kx$
   e) $y:\dfrac{1}{x} = k$

6. Look and read:

The volume of a gas is inversely proportional to its pressure.
The smaller the volume, the higher the pressure.

Now make similar sentences about the following:

   a)   the density and pressure of a gas (for a constant temperature)
   b)   the volume and temperature of a gas (for a constant pressure)
   c)   the velocity of a falling body and the time it has been falling
   d)   acceleration and mass for a constant force
   e)   the electrical resistance of a wire and its cross-sectional area
   f)   the circumference of a circle and its diameter
   g)   the volume of a cylinder and its cross-sectional area and height

Section 3   Reading

7. Read this:

The ratio of two quantities is the magnitude of one quantity relative to the other. Division of the quantity a by the quantity b gives the ratio $\dfrac{a}{b}$, which can also be written as a:b and is read as 'the ratio of a to b'. For example, the ratio of boys to girls in a particular school is 3:2. If the school has 250 pupils, then we can see that $\frac{3}{5}$ of these are boys and $\frac{2}{5}$ girls, i.e. there are 150 boys and 100 girls.

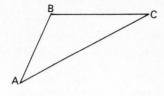

Relative sizes of more than two quantities may be expressed by ratio. For example, the ratio of AB:BC:AC in triangle ABC is 3:4:6. Hence we can see that AC is twice as long as AB. Any triangle which is similar to triangle ABC has sides in exactly the same ratio.

When the ratio of one pair of quantities is equal to the ratio of another pair of quantities, the two pairs are said to be in proportion. If we say that a,b,c,d, are in proportion, we mean that $\frac{a}{b} = \frac{c}{d}$. A property of this proportion is that the reciprocals are also in proportion. Moreover, the ratio of the numerators is equal to the ratio of the denominators.

Say whether the following statements are true or false. Correct the false statements.

a) In a test a student scored 30 out of 100. This gives a ratio of correct answers to incorrect answers of 3:10.

b) The length of the shortest side of a triangle similar to triangle ABC is 12 cm. The other sides are therefore 16 and 24 cm long.

c) Any triangle with sides in the ratio 2:3:6 is a right-angled triangle.

d) We may use ratios to express the relationship between any number of quantities.

e) If a, b, c, d are in proportion, then ad = bc.

8. Complete these exercises:

The last two sentences of the reading passage contain two properties of the proportion $\frac{a}{b} = \frac{c}{d}$. Put these two properties into mathematical form, and number them 1 and 2.
In the calculations which follow, there are four more properties of the proportion. Complete the proof by adding expressions from the lists.

From properties 2 and 3, we have   But
Adding 1 to each side, we obtain   Therefore
From properties 2 and 5, we have   Given
Similarly   Substituting, we obtain
From properties 2 and 4, we have

a) $\frac{a}{b} = \frac{c}{d}$ (......)

b) ...... $\frac{a}{b} + 1 = \frac{c}{d} + 1$

62

c) ...... $\dfrac{b}{b} = 1$ and $\dfrac{d}{d} = 1$

d) ...... $\dfrac{a+b}{b} = \dfrac{c+d}{d}$ (Property 3)

e) ...... $\dfrac{a-b}{b} = \dfrac{c-d}{d}$ (Property 4)

f) ...... $\dfrac{a+b}{c+d} = \dfrac{b}{d}$

g) ...... $\dfrac{a-b}{c-d} = \dfrac{b}{d}$

h) ...... $\dfrac{a+b}{c+d} = \dfrac{a-b}{c-d}$ (Property 5)

i) ...... $\dfrac{a+b}{a-b} = \dfrac{c+d}{c-d}$ (Property 6)

Section 4   Listening

Using percentages in statistics

9. **Listen to the passage and write down these words in the order in which you hear them:**

passed	respectively	while
whereas	relative	higher
improvement	yesterday	meat

10. **Copy the following diagrams, which illustrate the statistics in the five examples:**

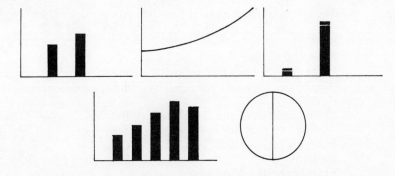

The five examples given in the passage may be summarised as: (1) bread; (2) wages; (3) exam; (4) restaurant; (5) meat.

a) Put the correct title under each diagram.
b) Label the diagrams as accurately as possible using the figures given in the passage.

11. Each of the examples given in the passage illustrates one of the following general principles. Decide which example illustrates which principle:

  a) Two successive increases of x % do not give an increase of 2x %.
  b) It is misleading to compare percentages if they are not related to the same quantity.
  c) Statistics should be based on a sufficiently large number of examples.
  d) Changes in quantities which are percentages should not be expressed in percentages.
  e) Choice of starting point for comparisons is important.

Think of other examples which illustrate the same general principles.

12. PUZZLE

A man has a litre of water and a litre of milk. He takes a glassful of the water and adds it to the milk. Then he takes a glassful of this milk-and-water mixture and adds it to the water. Is there more milk in the water than water in the milk, or vice versa?

Unit C   Revision

1. Look and read:

The volume of a right rectangular prism is found by using the formula $l \times h \times w$, where l = length, h = height and w = width.

Now make similar sentences using this table:

	Volume	Surface Area
Sphere	$\frac{4}{3}\pi r^3$	$4\pi r^2$
Cylinder	$\pi r^2 h$	$2\pi r(r+h)$
Cone	$\frac{1}{3}\pi r^2 h$	$\pi r(r+\sqrt{r^2+h^2})$ or $\pi r(r+l)$

2. Look and read:

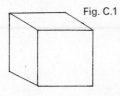

Fig. C.1

The right rectangular prism in figure C.1 is a cube.

Eliminating h and w from the above formula gives volume = l^3.

Fig. C.2

 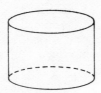

The radii of the sphere and of the bases of the cylinder and cone in Figure C.2 are all equal. In addition the height of the cylinder and the cone are both equal to the diameter of the sphere. Therefore h can be eliminated from the above formulae.

Now make correct sentences from the table:

Eliminating h from the formula for the

$$\left\{\begin{array}{l}\text{volume}\\\text{surface area}\end{array}\right\} \text{ of the } \left\{\begin{array}{l}\text{cylinder}\\\text{cone}\end{array}\right\} \text{ gives } \left\{\begin{array}{l}6\pi r^2\\3\cdot24\pi r^2\\\frac{2}{3}\pi r^3\\2\pi r^3\end{array}\right\}$$

3. **Now answer these questions about the solids in exercise 2:**

    a) Which solid has the greatest volume?
    b) Of the sphere and the cone, which solid is larger and by how much?
    c) Compare the cylinder and the cone in the same way.
    d) Which solids may be inscribed in which other solids?
    e) What is the ratio of the volumes of the three solids?
    f) Compare the surface area of the sphere with the area of the curved surface of the cylinder.

4. **Look and read:**

Fig. C.3

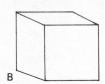

The length of the diagonal of one face of cube A in Figure C.3 is equal to the diameter of the sphere in Figure C.2. The length of one edge of cube B is also equal to the diameter of the sphere.
Cube B cannot be inscribed in the sphere, as the sphere is not large enough.

Now make sentences using the following words:

    a) Cube A ...... inscribed ...... sphere ...... too small.
    b) Cube A ...... circumscribed ...... sphere ...... large enough.
    c) The cylinder ...... inscribed ...... cube A ......
    d) The cylinder ...... inscribed ...... cube B ......
    e) Cube A ...... circumscribed ...... cone ......
    f) The cone ...... inscribed ...... cube B ......

5. **Look and read:**

Fig. C.4

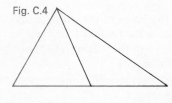

A straight line drawn from any vertex of a triangle to the mid-point of the opposite side is known as a median of a triangle. The three medians of a triangle are *concurrent*. The point where they meet is inside the triangle and is called the centroid, or the centre of gravity of the triangle.

From the information given in the above paragraph, complete the first row of the table below:

Figure	Name of line	Concurrent?	Point of concurrency	
			Always inside?	Name
C.4				
C.5	angle bisector	✓	✓	centre of the inscribed circle
C.6	perpendicular bisector	✓	×	centre of the circumscribed circle
C.7	altitude	✓	×	orthocentre

With the above paragraph as a model, write three paragraphs about the following figures using the information given in the table:

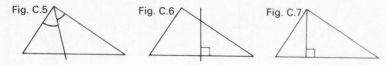

Fig. C.5   Fig. C.6   Fig. C.7

6. Read this:

To find the orthocentre of a triangle, it is sufficient to construct two altitudes. Their point of intersection is the orthocentre.

Say how to find the following:

a) the centre of gravity of a triangle
b) the centre of the inscribed circle of a triangle
c) the centre of the inscribed circle of a rhombus
d) the centre of the circumscribed circle of a triangle
e) the centre of the circumscribed circle of a rectangle
f) the centres of the escribed circles of a triangle (see Figure C.8).

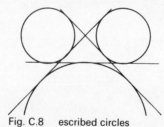

Fig. C.8   escribed circles

7. Draw a bar graph illustrating the information given in the following passage:

The largest ocean in the world is the Pacific, with an area of almost 170 million square kilometres, slightly more than twice as big as the Atlantic. The sizes of the Atlantic, Indian and Arctic Oceans are approximately in the ratio 6:5:1. The Arctic Ocean is over five times as large as the Mediterranean Sea, which is about 260 000 square kilometres bigger than the South China Sea and the Bering Sea. Of these last two seas, the former is slightly larger than the latter.

8. Write a paragraph about the information given in the following graph:

(**Note**: 'per capita income' refers to the arithmetic mean of the money earned by all the people in a country in one year.)

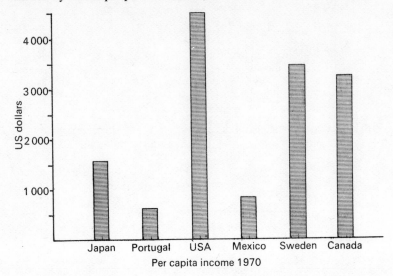

Unit 10 Measurement 4 Frequency, Probability, Tendency

Section 1 Presentation

1. Look and read:

A coin is tossed. There are two possible *outcomes*.

The outcome may be *heads*, or it may be *tails*. The outcome will certainly be one or the other, and it will certainly not be both.

Complete the following sentences, using the words *and*, *or*, *will* and *may*:

a) A dice is thrown.
The outcome _____ be odd, _____ it _____ be even. The outcome _____ certainly be one or the other, _____ it _____ certainly not be both.

1, 2, 3, 4, 5,
6, 7, 8, 9, 10.

b) A number is chosen at random from the first ten integers.
The outcome _____ be even. It _____ be a prime number. It _____ be less than five. It _____ be all of these _____ it _____ be none of them.

Complete the following sentences, using *both*, *neither*, *all* and *none*:

c) This book is opened at random between pages 10 and 100.
The outcome may be a page with a diagram, or it may be a page with a reading passage. It may be _____ or it may be _____. It will certainly have _____ of the following: words, sentences and numbers.

4, 6, 8, 9, 10 ......

d) A non-prime number is chosen at random from the first fifty integers.
The outcome may be a number divisible by 2, 3 or 5. It may be divisible by _____ of them or by _____ of them.

69

2. Read this:

> A coin is tossed. The outcome is either heads or tails, but not both. One result *excludes* the other. Heads and tails are *mutually exclusive events*. No other results are possible, therefore the two outcomes heads and tails are *exhaustive*.

Say whether the following events are mutually exclusive and/or exhaustive. Use the words *neither ...... nor, both ...... and* or *but* in your answers.

> *Example:* The events *heads* and *tails* are both mutually exclusive and exhaustive.

a) A card is drawn at random from a pack of playing cards.
   i) red/black
   ii) clubs/diamonds/hearts
   iii) red/picture
   iv) higher than 10/lower than 10
b) A number is chosen at random from the set of natural numbers.
   i) odd/even
   ii) prime/even
   iii) multiples of 2/multiples of 3

3. Read this:

> A dice is thrown once.
> The two outcomes, odd or even, are *equally likely*.
> The chances of throwing an odd number are $\frac{1}{2}$ or 1 in 2.
> The chances of throwing neither an odd nor an even number are 0.
> The chances of throwing either an odd or an even number are 1.

A card is drawn from a pack of playing cards. What are the chances of drawing the following cards?

a) a red card
b) the five of hearts
c) a picture card
d) a club
e) neither a red card nor a black card
f) either a club or a heart
g) either a picture card or a non-picture card

Section 2   Development

4. Read this:

> The outcomes of twenty *consecutive* tosses of a coin are as follows: HTTTHTTHHHHHHHTHTTTTT. The first outcome, heads, is followed by a *run* of three tails. This is followed by a single head. Describe the rest of the *sequence*.

70

5. Look and read:

5 36 18 27 13 28 6 24 25 21 34 6 21 14 2 14 5 36 33 22 2 7 4 9 10 9 3 14 14 12 28 7 25 11 33 23 4 24 3 23 13 21 12 35 1 25 20 15 25 5 6 22 25 28 10 27 34 30 13 2 24 28 29 29 21 6 20 26 35 27 25 21 3 35 34 10 24 29 20 23 32 1 4 6 31 10 27 31 21 34 19 13 12 32 24 6 4 23 23 34 36

The above table shows the outcomes of 100 consecutive random selections of numbers from the first 36 integers.

Look at these sentences:

- There are no runs of three identical numbers.
  Runs of three identical numbers never occur.
- There are a few cases where three consecutive numbers form an arithmetical series.
  Three consecutive numbers occasionally form an arithmetical series.

Now change these sentences to sentences of the second kind using the words *often, rarely, never, occasionally*.

a) There are no runs of five single-digit numbers.
b) There are many runs of five two-digit numbers.
c) There are very few runs of more than two single-digit numbers.
d) There are a few cases where a number is followed by its own square root.
e) There are no cases of the outcome 8.

6. Look and read:

a) Although runs of three identical numbers never occur in this sample, such an outcome could occur if, for example, 500 further random selections were made. In fact, a run of three identical numbers is far from improbable.
   The above paragraph is summarised in this table.

Event	First 100	Next 500	
		Possible?	Probable?
a)	No	Yes	Yes

Copy the table and complete it for these other cases:

b) Although there is a run of seven numbers $\leqslant 10$, such an event is quite rare, and is unlikely to be repeated in the next 500 selections.
c) Although there are no cases of the outcome 8, it is extremely unlikely that such an absence would continue over the next 500 selections.

d) Although ten successive integers could in theory occur in ascending or descending order, such an event is highly unlikely.
e) Although 23 occurs more often than 18 in the first 100 selections, the opposite is just as likely to occur over the next 500 selections.

Section 3   Reading

7. Read this:

How many different five digit numbers can be formed from the digits 1, 2, 3, 4 and 5? The first position in the number may be occupied by any one of the five digits, i.e. there are five possibilities for the first digit of the number. When this position has been filled, only four possibilities remain for the second position. For the third position, only three possibilities remain, for the fourth only two, and for the final position there is no choice. Hence the total of different numbers that can be formed from the five digits is $5 \times 4 \times 3 \times 2 \times 1 = 120$. In general, for n different digits, the total number of possible arrangements is $n \times (n-1) \times \ldots\ldots \times 3 \times 2 \times 1$. Such products occur frequently in mathematics and are denoted by n!, which is read as n factorial.

How many different five-digit numbers can be formed from the digits 1, 2, 3, 4 and 5 if we can use each digit more than once? The first position in the number may be occupied by any one of five possibilities and so may the second, and the third, and the fourth and fifth. So the total number of possibilities is $5 \times 5 \times 5 \times 5 \times 5 = 3\,125$.

The first case is called *sampling without replacement*, because when an element has been selected once, it cannot be selected again. The second case is called *sampling with replacement*, because each element may be chosen any number of times.

Say whether the following statements are true or false. Correct the false statements.

a) 5! is read as 5 factorial.
b) The number of different five-digit numbers that can be formed from the digits 5, 6, 7, 8, and 9 is 9!
c) In both examples, the probability of choosing 5 as the first digit is 1 in 5.
d) The number 11111 is possible in neither example.
e) In the first example, the numbers 12345 and 54321are equally likely.

8. Look at this example:

Without replacement, there are 120 ways in which the five digits 1, 2, 3, 4 and 5 may be arranged.

Now make similar sentences about the following:

a) The letters ABCDEF without replacement.
b) The letters ABCDEF with replacement.
c) Four people in a queue.
d) The digits 1, 2, 3 and 4 without replacement, if the first digit of the arrangement is 4.

Section 4 Listening

Pascal's triangle

9. **Listen to the passage and write down the word in each of the following pairs which occurs in the passage:**

simple/single
also are/and so are
a last/the last
producers/produces
called/call
sum/some
these/threes

Fig. 10.1

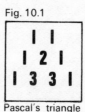

Pascal's triangle

10. **Write down the next three rows of Pascal's triangle:**

Using the triangle, state the chances of the following events when throwing six dice.

a) all odd
b) either all odd or all even
c) an equal number of odd and even
d) five odd and one even or vice versa.

11. PUZZLE

Comment on the following:

Question: Two tennis players, Alan and Bob, are equally good at tennis. If they play two matches, what are the chances that Alan will win at least one match?
Answer: There are three possibilities: Alan loses both matches; Alan wins one match; Alan wins both matches. Out of the three possibilities two satisfy the requirement that he wins at least one match, and the probability of this event is, therefore, $\frac{2}{3}$ or 2 in 3.

Unit 11   Process 4   Method

Section 1   Presentation

1. Look and read:

Transformations

In the following examples, △ABC is congruent with △XYZ.

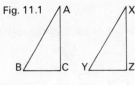

In Figure 11.1 triangle XYZ is obtained by *translating* triangle ABC to the right.

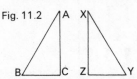

In Figure 11.2 triangle XYZ is obtained by *reflecting* triangle ABC.

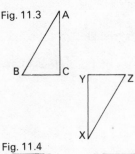

In Figure 11.3 triangle XYZ is obtained by *rotating* triangle ABC through 180°.

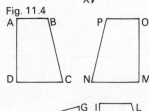

Say how the trapeziums in Figure 11.4 are obtained from ABCD.

2. Read this:

In Figure 11.3 it is possible to rotate triangle ABC through 180° *into* triangle XYZ.

Write similar sentences about the transformation of Figures 11.1, 11.2 and 11.4.

3. Look and read:

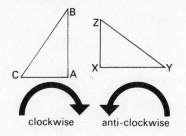

clockwise   anti-clockwise

One method of transforming $\triangle ABC$ into $\triangle XYZ$ is to rotate it clockwise through 90°. An alternative method is to rotate it anti-clockwise through 270°.

Now write correct sentences from the table:

One An alternative Another	method of	multiplying two large numbers solving a triangle solving quadratic equations

	is to	draw a graph. use a calculator. factorise the equation. use the sine and cosine laws. use a formula. use logarithms. construct it. complete the square. use a slide rule.

Section 2   Development

4. Read this:

Axioms

Some mathematical *laws* are accepted without *proof*. These fundamental laws are known as *axioms*. In algebra, for example, there are the following axioms.

1) $x+y = y+x$
2) $xy = yx$
3) $x+(y+z) = (x+y)+z$
4) $x(yz) = (xy)z$
5) $x(y+z) = xy+xz$
6) $x+0 = x$
7) $1x = x$
8) For every real number x, there is a real number y such that $x+y = 0$.
9) For every non-zero real number x, there is a real number y such that $xy = 1$.

75

- It is axiomatic that if $x+y = a$, then $y+x = a$.

Write similar sentences about the other axioms.

5. **Read this:**

- Axiom 1 is illustrated by the following example: $2+3 = 5$; $3+2 = 5$.

Write similar sentences using these examples:

a) $1 \times 22 = 22$
b) 5 is a non-zero real number; $1/5$ is a real number
c) $5 \times 7 = 35$; $7 \times 5 = 35$
d) $3 \times (11+19) = 3 \times 30 = 90$; $(3 \times 11)+(3 \times 19) = 33+57 = 90$.
e) $3+0 = 3$
f) $3+(4+5) = 3+9 = 12$; $(3+4)+5 = 7+5 = 12$
g) 5 is a real number: -5 is a real number
h) $2 \times (3 \times 4) = 2 \times 12 = 24$; $(2 \times 3) \times 4 = 6 \times 4 = 24$

Section 3   Reading

6. **Read this:**

Theorems

From the axioms given in the last section we can prove all the laws of algebra. Laws which are not axioms are called *theorems*. There are various methods of proving theorems. Some examples are given here.

1   *Proof by deduction*
Theorem (1)   To prove: If $a+b = a+c$, then $b = c$.

Proof
By axiom (8), there is a number y such that $y+a = 0$.
Adding y to both sides gives $y+(a + b) = y+(a+c)$.
By axiom (3), we have $(y+a)+b = (y+a)+c$.
But $y+a = 0$, therefore $0+b = 0+c$.
By axiom (6), $0+b = b$ and $0+c = c$.
Therefore $b = c$.

2 and **3**   *Proofs by elimination and contradiction*
Thereom (2)   To prove: Given a and b, there is one and only one x such that $a+x = b$.

Proof:
There are three possibilities: more than one x, less than one x, exactly one x. If we eliminate two of these possibilities, then the third must be true. We can divide the proof into three parts.

a)   To prove that there is not less than one x such that $a+x = b$.
    By axiom (8), there is a number y such that $a+y = 0$.

Let $x = y + b$.

Then $a + x = a + (y + b)$.

By axiom (3) $a + x = (a + y) + b$.

But $a + y = 0$, therefore $a + x = 0 + b$.

By axiom (6), $a + x = b$.

Hence there is not less than one x such that $a + x = b$.

b) To prove there is not more than one x such that $a + x = b$.

We can prove this by using proof by contradiction, i.e. we assume the opposite of what we are trying to prove and show that this leads to a contradiction.

Assume that there are several different x, such that $x_1 \neq x_2 \neq x_3$, etc.

and $a + x_1 = b$, $a + x_2 = b$, $a + x_3 = b$, etc.

i.e. $a + x_1 = a + x_2 = a + x_3$, etc.

By theorem (1), we have $x_1 = x_2 = x_3$, etc., but this contradicts our assumption that $x_1 \neq x_2 \neq x_3$, etc.

Therefore there is not more than one x such that $a + x = b$.

c) Two possibilities have been eliminated, therefore the only remaining possibility is that there is one and only one x such that $a + x = b$.

Say whether the following statements are true or false. Correct the false statements.

a) Theorems are fundamental laws.

b) There are not less than three ways of proving theorems.

c) All of the symbols a, b, x, y used in the above proofs refer to real numbers.

d) All of the symbols a, b, x, y used in the above proofs refer to non-negative numbers.

e) The three kinds of proof shown are mutually exclusive.

7. Look and read:

Here are the proofs of two more theorems. Complete them by inserting the correct words and say what kinds of proof are used.

a) Theorem (3):   If $ab = ac$ and $a \neq 0$, then $b = c$.

By ...... there is a number y ...... $ya = 1$.

_____ both sides by y, ... $y(ab) = y(ac)$.

By ...... $(ya)b = (ya)c$.

But $ya = 1$, _____ $1b = 1c$.

By ......, $1b = b$ and $1c = c$.

_____ $b = c$.

b) Theorem (4):   Given $a \neq 0$ and b, there is one and only one x such that $ax = b$.

There are three possibilities, ... one x, ... one x, _____ one x.

   i) By ...... there is a number y ...... $ya = 1$.

_____ x = yb.
_____ ax = a(yb).
By ......, we have ax = (ay)b.
_____ ay = 1, _____ ax = 1b.
By ...... ax = b.
_____ there is ......

ii) _____ $ax_1 = b, ax_2 = b, ax_3 = b$ etc.
and $x_1 \neq x_2 \neq x_3$ etc.
_____ $ax_1 = ax_2 = ax_3$ etc.
By ......, we have $x_1 = x_2 = x_3$ etc.
But this _____ our _____ that $x_1 \neq x_2 \neq x_3$ etc.
_____ there is ......

iii) _____ the only _____ possibility is that there is ......

Section 4   Listening

Simultaneous equations

8. Listen to the passage and write down the correct equations to complete the following as you hear them:

a) (1) $x + y = 7$
(2) ......
Subtracting (2) from (1) gives ......
Substituting in (1) gives ......

b) (1) $3x + y$ ......
(2) ......
From (1) we have ......
Substituting in (2) gives ......
This gives ......
Factorising, ...
Therefore, $x = $ ...... or ......
By substitution, $y = $ ...... or ......

9. Look at these:

$x + 2y = 9$
$4x + y = 8$
These are simultaneous linear equations in two unknowns.

Give full descriptions of the following equations:

a) $x + 2y + z = 11$
$4x + y + 3z = 14$
$2x + 3y + 4z = 22$

b) $5x + y = 13$
$2xy = 12$

10. Complete the following sentences:

a) Simultaneous equations contain ...... unknowns.
b) Simultaneous equations in two unknowns can be solved by ......
c) To solve simultaneous equations in n unknowns, it is necessary ......
d) Linear equations do not _____ the use of powers.
e) The equation $3x^2 - 11x + 6 = 0$ can be easily solved by _____

11. PUZZLE:

a) Using six matches, construct four congruent equilateral triangles, each with sides one match long.
b) Can you make this animal look in the opposite direction by changing the position of just two matches?

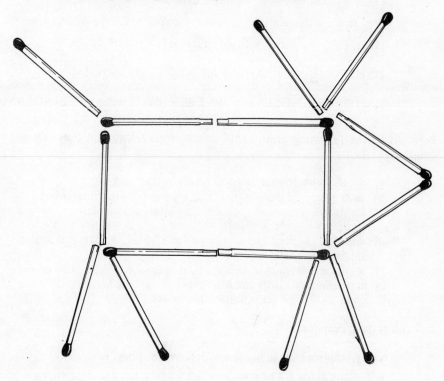

Unit 12　Consolidation

1. Look at these words:

circles, spheres, solids, geometrical figures, triangles, plane figures, cylinders

These words may be divided into three levels of generality, as follows:

LEVEL 1 (most general):　geometrical figures
LEVEL 2　　　　　　　 :　solids; plane figures
LEVEL 3 (most specific):　spheres, cylinders; triangles, circles

We can show the relation between the words by using a tree diagram:

LEVEL 1　　　　　GEOMETRICAL FIGURES

LEVEL 2　　　PLANE FIGURES　　　SOLIDS

LEVEL 3　TRIANGLES　CIRCLES　SPHERES　CYLINDERS

Now divide the following groups into levels of generality and draw a tree diagram for them:

a)　sine, cosine, trigonometrical ratios, ratios, tangent
b)　inch, metre, SI units, non-SI units, pound, units, kilogramme
c)　hardware, programs, computers, input units, software, output units
d)　speed, velocity, mass, vector quantities, scalar quantities, quantities
e)　addition, subtraction, arithmetical operations, raising to a power
f)　mathematical instruments, setsquares, protractors, 45° set squares, 60°/30° set squares, calculators

2. Look at these examples:

- A triangle is a plane figure which has three sides.
- A plane figure is a geometrical figure which has two dimensions.
- A calculator is a mathematical instrument which enables us to make rapid calculations.

The idea of 'more general' and 'more specific' can be used in describing.

The sentences are formed like this:

| SPECIFIC WORD | + | Verb 'to be' | + | GENERAL WORD | + | which...... |

Now describe the following in the same way:

a) Vector quantities
b) A setsquare
c) The sine ratio
d) A prime number
e) The diameter of a circle
f) A rational number
g) Axioms

3. Look at these examples:

- Descartes (1596–1650) developed the Cartesian co-ordinate system.
  The Cartesian co-ordinate system was developed by Descartes (1596–1650).
- Pascal (1623–1662) invented a calculating machine.
  A calculating machine was invented by Pascal (1623–1662).

Now make similar statements about the following:

a) Euclid (365?–300?BC)/wrote/the 'Elements'.
b) Lobachevsky (1792–1856) and Bolyai (1802–1860)/discovered/ non-Euclidean geometries.
c) Newton (1642–1727) and Leibniz (1646–1716)/developed/ calculus.
d) Cauchy (1789–1857)/developed/a theory of limits.
e) Boole (1815–1864)/originated/mathematical logic.

4. Now look at the following paragraph:

The Cartesian co-ordinate system is a device which represents diagrammatically any point in a plane. It is named after René Descartes (1596–1650) who first wrote about it in 1637 in *La Géométrie*.

Write similar paragraphs using the following information:

	Person	Dates	Contribution to mathematics	Book
a)	Leonard Euler	1707–1783	Discovery of Euler's theorem (see Unit 4 paragraph 5)	Letter to Goldbach (1750)
b)	John Napier	1550–1617	Development of Napierian logarithms (logarithms to the base e)	*Mirifici Logarithmorum Canonis Descriptio* (1614)
c)	Blaise Pascal	1623–1662	Invention of Pascal's triangle (see Unit 10 paragraph 9)	*Traité du triangle arithmétique* (1654)
d)	Johannes Kepler	1571–1630	Discovery of Kepler's Laws (mathematical laws which govern the motions of the planets)	*Epitome astronomiae Copernicanae* (1617)

5. Read this:

The Möbius strip

The Möbius strip is a construction which has some very strange properties. It is named after Möbius (1790–1868) who first wrote about it in 1865 in a book called *Über die Bestimmung des Inhaltes eines Polyëders*.

The properties of the Möbius strip can most easily be discovered by observation. To do this, take a long strip of paper (Fig. 12.1). Now twist it once (Fig. 12.2). Finally stick one end of the paper to the other (Fig. 12.3).

Now you can try a few experiments with this Möbius strip. If you try to colour only *one* surface of the strip, you will find that it is impossible. Drawing a continuous line along the middle of the strip produces a line on both sides of the paper! By cutting along this line, we would expect to divide the strip into two halves, but in fact we form a longer, thinner strip which is still in one piece.

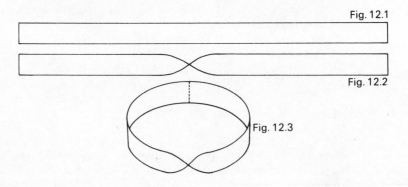

Fig. 12.1

Fig. 12.2

Fig. 12.3

a) Construct a Möbius strip and draw a line along its length approximately $\frac{1}{3}$ of the distance between one edge and the other. Cut along this line and describe the result.

b) Describe how to construct a similar strip with two twists and describe the results of similar experiments to those mentioned in the text and in question 8(a).

c) Do the same for a strip with three twists.

6. Say whether the following statements are true or false. Correct the false statements.

a) A Möbius strip has only one surface.

b) A Möbius strip has two edges.

c) A Möbius strip is a three-dimensional figure.

d) Observing a Möbius strip is the easiest way to discover its properties.

e) Sticking together the ends of a twisted rectangle produces a Möbius strip.

Listening

The Fibonacci series

7. Listen to the passage and write down the words from the following lists which occur in the passage:

sum	two	four
some	too	was
serious	for	were
series	form	

8. Now answer the following questions:

- a) When did Leonardo Fibonacci live?
- b) What is a recursive series?
- c) Give two examples of mathematical areas where recursive series have been important.
- d) Give an example of another science where recursive series have been important.
- e) Continue the series given in the passage beginning $\frac{1}{2}, \frac{1}{3}, \ldots\ldots$

9. Which of the following are recursive series?

- a) $-1, -1, -2, -3, -5, -8, -13, \ldots\ldots$
- b) $1, 1, 2, 2, 3, 3, 4, \ldots\ldots$
- c) $1, 3, 4, 7, 11, 18, \ldots\ldots$
- d) $x, x, 2x, 3x, 5x, 8x, \ldots\ldots$
- e) $\frac{1}{4}, \frac{1}{2}, \frac{3}{4}, 1, 1\frac{1}{4}, 1\frac{1}{2}, \ldots\ldots$

10. PUZZLE 1:

Say what is being described in each of the following sentences or paragraphs:

- a) This is a solid figure which is formed by rotating a right triangle through 360° around either of the sides adjacent to the right angle.
- b) This is a device which consists of two lines intersecting at right angles. It is particularly useful for giving visual representations of functions.
- c) This is a standard set of numbers which includes the square root of 2 and all natural numbers but not the square root of minus 1 nor minus 5.
- d) This is a method of expressing all numbers as the product of prime numbers.
- e) This is an object used in games. It is shaped like a polyhedron and has 8 vertices. The sum of numbers written on opposite faces is 7.

11. PUZZLE 2:

A large piece of paper 1 mm thick is cut in two and the two halves are then placed one on the other. These pieces are then cut in two and the two resulting halves placed on top of each other. The pile of paper is now 4 mm thick. How thick will it be after 50 such cuts?

Listening passages

Unit 1

Draw a horizontal line in the middle of the page. Label it AB. From B draw a perpendicular line shorter than AB. Label it BC. Join AC. From C draw a line parallel to AB; to the left draw CD so that CD = BA and to the right draw CE so that CE = CD. Join AD and BE. Produce AD to F so that DF = AD. Join EF. Join FB which intersects CD at G and AC at H.

Unit 2

Draw a large circle with centre 0. Draw a horizontal line AB which is a diameter of the circle. Draw a chord CD above AB and parallel to AB. Join AC and BD. On the arc CD, mark a point E so that the arc EC = arc ED. Join AE, BE and OE. CD intersects AE at F BE at G and OE at H. At E, draw a tangent to the circle. Label the tangent JEK. In the other semicircle, draw a chord LM parallel to AB. Join AM and BL, intersecting at N.

Unit 3

The number system

The set of positive and negative integers consists of all the natural numbers 1, 2, 3, 4, ..., plus the same numbers preceded by the minus sign, $-1, -2, -3, ...$ We can represent any of these numbers on the number line. We can also represent fractions of numbers, e.g. $1 \cdot 5$, $\frac{2}{3}$, $-3 \cdot 4$ etc., on the number line. The rational numbers are composed of both the integers (or whole numbers) and the non-integers (or fractions). All rational numbers may be represented as a fraction where both the denominator and the numerator are integers, whereas irrational numbers cannot be expressed in this way. Irrational numbers include numbers like π $(3 \cdot 14159)$, $\sqrt{2}$ $(1 \cdot 41421)$, $\sqrt[3]{5}$ $(1 \cdot 70997)$, and so on. All these numbers, both rational and irrational, make up the set of real numbers, and may be represented as points on a number line. Imaginary numbers, on the other hand, cannot be represented as points on a number line. They include numbers such as $\sqrt{-1}$, which is usually expressed by the symbol i. Finally, a complex number is a number which contains both a real number and an imaginary number, for example $6 + \sqrt{-4}$.

Unit 4

Vectors and scalars

A car is travelling north along a road at 60 km h^{-1}. We say that it has a velocity of 60 km h^{-1} north. It has a speed of 60 km h^{-1}. The second quantity, speed, is a scalar quantity, that is, it is a size, or magnitude. The first quantity, velocity, is a vector, that is, it has both magnitude and direction.

The magnitude of a quantity is usually expressed in relation to a standard unit of magnitude. For example, if the mass of a metal cylinder is 2 500 g, then it is 2 500 times the unit of mass, that is, the gramme. If the electrical power of a light bulb is 40 W, then it is 40 times the unit of power the watt. Both of these are scalar quantities.

If a certain town is 150 km from London, then the distance from London to the town is 150 000 times the unit of distance, the metre. Distance, again, is a scalar quantity. Now, if we want to know the exact location of the town, we also need to know the direction, that is, we need the vector quantity displacement, which consists of both distance and direction, 150 km north west.

Unit 5

Sets of numbers

Several sets of numbers are used frequently in mathematics and the use of standard abbreviations or symbols to refer to them enables us to save time and space. Capital letters are usually used for this notation. The set of natural numbers is denoted by N. Z denotes the set of all integers. R represents all real numbers and Q all rational numbers.

Note that Z^+ means all positive integers, R^- all negative real numbers, and so on. Z^+ differs from N in that Z^+ consists of all positive integers, whereas N consists of all non-negative integers and therefore includes the element zero in addition to the elements of Z^+. We can see from this that a set which contains only the element *zero* is not the same as the empty set, which contains no elements. Thus $N - Z^+ = \{0\}$, but $Z^+ \cap Z^- = \varnothing$.

Some other sets are also referred to by abbreviations. We use a capital U to refer to the universal set, while the empty set is denoted by a symbol which consists of a zero bisected by an oblique line. This may also be read as the null or void set. For example, if two sets, A and B are disjoint (that is, they have no elements in common), then $A \cap B = \varnothing$.

Unit 6
Stationary points
The gradient of a graph is the increase in y $\div$ the increase in x. Figs. 6.7, 6.8 and 6.9 show three straight-line graphs. In Fig. 6.7, the line of the graph is parallel to the x-axis, and we say that the gradient is zero. In Fig. 6.8, y increases as x increases, that is, the gradient is positive. In Fig. 6.9, y decreases as x increases, in other words the gradient is negative. In the graph of $y = x^2$ (Fig. 6.10), the gradient is first negative, becoming less and less steep as x approaches zero. Then, at the origin, the gradient becomes zero before turning and becoming positive.

A point on a graph where the gradient is zero is a stationary point. There are three kinds of stationary points: (a) when the gradient is positive before the point and negative after, the point is said to be a local maximum; (b) when the gradient is negative before the point and positive after, the point is said to be a local minimum; and (c) when the sign of the gradient is the same before the point as after, the point is said to be a point of inflexion. All maxima and minima are turning points. whereas a point of inflexion is not.

Unit 7
Approximate values
The value of π may be calculated to any required degree of accuracy. Correct to four decimal places, its value is 3·1416. This value is said to be correct to five significant figures.

If the population of a city is 346 268, then we may say that the population is approximately 350 000. This approximation is said to be correct to two significant figures.

In this last case the approximate value exceeds the true value by 3 732. This difference is known as the absolute error or the true error.

Another important value is *relative* error. We can use the formula $\dfrac{\text{absolute error}}{\text{true value}}$ to find the relative error. In this case, we have

$$\frac{3\,732}{346\,268} = 0{\cdot}0108.$$

Note that the calculation of the relative error, 0·0108, is accurate to three significant figures.

We can also express the relative error as a percentage, in this case 1·08 %.

If we multiply one approximate value by another, the number of significant figures in the product is generally less than in the multiplier and multiplicand. For example, the product of 3·76, accurate to 3 significant figures and 2·012, accurate to four significant figures, is 7·56512, which is only accurate to two significant figures, 7·6, as the true answer may be anywhere between 7·553 and 7·577.

Unit 8

Multiplication of vectors by scalars

The effect of multiplying a vector quantity by a scalar quantity is to alter the magnitude but leave the direction unchanged. Take, for example, the vector $\overline{OA}$ in Fig. 8.6. The length and direction of OA may be specified either by using rectangular co-ordinates ((x,y) in this example) or polar co-ordinates ((r,θ) in this example).

If we multiply $\overline{OA}$ by a scalar quantity, say 2, then the length of OA becomes 2r, while the angle θ does not change. In the same way, the rectangular co-ordinates become (2x, 2y).

Multiplying a vector by a negative quantity again does not change the angle but the process changes the direction of the vector. Thus, multiplying $\overline{OA}$ in Fig. 8.6 by -1 gives a vector $\overline{OP}$ equal in length to $\overline{OA}$ but pointing in the opposite direction.

Unit 9

Using percentages in statistics

Percentages are often used in statistics to represent one quantity relative to another. But it is easy to use percentages in a misleading way. Here are some examples.

1 The price of bread for the last five years has been, respectively, 10p, 14p, 19p, 24p, 21p. The person who buys bread can say that the price of bread has increased by 110 % in five years, whereas the person who sells bread can say that the price of bread has dropped by $12\frac{1}{2}$ % in twelve months.

2 One man's wages go up by 50 %, while another man's only go up by 10 %. But this does not mean that the first man is therefore richer. If the first man's wages were £20 a week, increasing to £30. and the second man's wages were £150 a week, increasing to £165, then the second man has had a larger increase than the first.

3 Last year 40 % of students in a school passed the English exam; this year 50 % passed. We can say either that the results are 10 % better than last year or that there has been a 25 % improvement.

4 A certain restaurant had only two customers yesterday. After the meal, one customer was ill. It is true to say that 50 % of the people who ate at the restaurant were ill.

5 The price of meat has risen by 10 % every year for the last ten years. But the price is not now 100 % higher than ten years ago, but 159 % higher.

Unit 10

Pascal's triangle

A single dice is thrown. There are two possible outcomes – odd or even. The chances of throwing an odd number are 1 in 2, and so are the chances of throwing an even number. Supposing now that two dice are thrown. There are now three possible outcomes: both odd, both even, or one odd and one even. But the last result, one odd and one even, can occur in two different ways, either the first dice odd and the second dice even, or the first dice even and the second dice odd. So the chances of throwing both odd are 1 in 4, of throwing both even

1 in 4, and of throwing one odd and one even 2 in 4. Throwing three dice produces eight different possibilities: EEE, EEO, EOO, EOE, OEE, OEO, OOE and OOO. Thus the probabilities are as follows: all evens, 1 in 8; all odds 1 in 8; two odds and one even, 3 in 8; two evens and one odd, 3 in 8.

These results can be calculated by using a device called Pascal's triangle, the first three rows of which are shown in Fig. 10.1. This triangle can easily be formed. The first and last figure in each row is one. Every other figure is the sum of the two figures above it. Thus, in the second row, 2 is the sum of 1 and 1. In the third row, both threes are formed by adding 2 and 1. The total of all figures in each row gives the total number of possibilities for that row. Thus the third row has $1 + 3 + 3 + 1$ possibilities.

Unit 11

Simultaneous equations

Consider the two equations $x + y = 7$ and $2x + y = 10$. Each of the equations contains the same two unknowns, x and y. Such equations are known as simultaneous equations. This example can be solved easily. Subtracting $x + y = 7$ from $2x + y = 10$ gives $x = 3$. Substituting $x = 3$ in $x + y = 7$ gives $y = 4$. Thus the solution of the simultaneous equations is $x = 3$, $y = 4$. In general. we find the value of one unknown by elimination of the other from the equations. With simultaneous equations in three unknowns, it is necessary to eliminate two unknowns; with four unknowns. three unknowns must be eliminated, and so on.

These equations are simple equations which do not involve the use of powers (e.g. x^2, y^3, y^4, etc.). They are therefore known as simultaneous linear equations or simultaneous first degree equations.

Simultaneous second degree equations involve the squares of one or more of the unknowns or the product of two unknowns. For example, consider the equations $3x + y = 11$ and $xy = 6$. From the first equation we have $y = 11 - 3x$. Substituting this in the second equation gives $x(11 - 3x) = 6$. This produces the quadratic equation $3x^2 - 11x + 6 = 0$, which factorises into $(3x - 2)(x - 3) = 0$, giving values of x at $+3$ and $+\frac{2}{3}$. Substituting these values in one of the first equations gives the complete solution of $y = 2$, $x = 3$, or $y = 9$, $x = \frac{2}{3}$.

Unit 12

The Fibonacci series

In earlier units of this book we have seen examples of both arithmetical and geometrical series. Another kind of series is a recursive series, in which each term of the series is the sum of the two preceding terms. The earliest example of a recursive series was the Fibonacci series, named after Leonardo Fibonacci, who devised the series and wrote about it in 1202 in a book called *Liber abbaci* (or 'The book of calculations'). The Fibonacci series begins 1, 1, 2, 3, 5, 8, 13, 21, 34 .... Fibonacci and other recursive series have been important in the study of mathematics, for example in the study of divisibility and prime numbers. They have also had applications in other sciences, for example biology, where the series $\frac{1}{2}, \frac{1}{3}, \frac{2}{5}, \frac{3}{8}, \frac{5}{13}, \ldots\ldots$ (in which both numerators and denominators form Fibonacci series) can be used in describing the arrangements of leaves on some plants.

Appendix I

Pronunciation of letters of the alphabet

English alphabet

a	/eɪ/	g	/dʒiː/	m	/em/	t	/tiː/
b	/biː/	h	/eɪtʃ/	n	/en/	u	/juː/
c	/siː/	i	/aɪ/	o	/əʊ/	v	/viː/
d	/diː/	j	/dʒeɪ/	p	/piː/	w	/ˈdʌbljuː/
e	/iː/	k	/keɪ/	q	/kjuː/	x	/eks/
f	/ef/	l	/el/	r	/ɑːʳ/	y	/waɪ/
				s	/es/	z	/zed/ AmE /ziː/

Greek alphabet

Letters Capital	Small	Name	Pron.	Letters Capital	Small	Name	Pron.
A	α	alpha	/ˈælfə/	N	ν	nu	/njuː/
B	β	beta	/ˈbiːtə/	Ξ	ξ	xi	/ksaɪ/
Γ	γ	gamma	/ˈgæmə/	O	o	omicron	/ˈəʊmɪkrən/
Δ	δ	delta	/ˈdeltə/	Π	π	pi	/paɪ/
E	ε	epsilon	/ˈepsɪlən/	P	ρ	rho	/rəʊ/
Z	ζ	zeta	/ˈziːtə/	Σ	σ, ς	sigma	/ˈsɪgmə/
H	η	eta	/ˈiːtə/	T	τ	tau	/taʊ/
Θ	θ	theta	/ˈθiːtə/	Y	υ	upsilon	/ˈjʊpsɪlən/
I	ι	iota	/aɪˈəʊtə/	Φ	φ	phi	/faɪ/
K	κ	kappa	/ˈkæpə/	X	χ	chi	/kaɪ/
Λ	λ	lambda	/ˈlæmdə/	Ψ	ψ	psi	/psaɪ/
M	μ	mu	/mjuː/	Ω	ω	omega	/ˈəʊmɪgə/

Appendix II

Pronunciation of some common mathematical expressions

Individual mathematicians often have their own way of pronouncing mathematical expressions and in many cases there is no generally accepted 'correct' pronunciation.

Distinctions made in writing are often not made explicit in speech; thus the sounds fx /'ef 'eks/ may be interpreted as any of: fx, f(x), f_x, FX, $\overline{FX}$. The difference is usually made clear by the context; it is only when confusion may occur, or where he wishes to emphasise the point, that the mathematician will use the longer forms: f multiplied by x, the function of x, f subscript x, line FX, vector FX.

Similarly, a mathematician is unlikely to make any distinction in speech (except sometimes a difference in intonation or length of pauses) between pairs such as the following:

$$x+(y+z) \quad \text{and} \quad (x+y)+z$$
$$\sqrt{ax+b} \quad \text{and} \quad \sqrt{(ax+b)}$$
$$a^n-1 \quad \text{and} \quad a^{n-1}$$

The most common pronunciations are given in the list below. In general, the *shortest* versions are preferred (unless greater precision is necessary).

$x+1$	x plus one
$x-1$	x minus one
$x\pm1$	x plus or minus one
xy	xy / x multiplied by y
$(x-y)(x+y)$	x minus y, x plus y
$\dfrac{x}{y}$	x over y
$x=5$	x equals 5 / x is equal to 5
$x\equiv y$	x is equivalent to y / x is identical with y
$x>y$	x is greater than y
$x\geq y$	x is greater than or equal to y
$x<y$	x is less than y
$0<x<1$	zero is less than x is less than 1
$0\leq x\leq 1$	zero is less than or equal to x is less than or equal to 1
x^2	x squared
x^3	x cubed
x^4	x to the fourth / x to the power four
x^n	x to the n /x to the nth/ x to the power n
x^{-n}	x to the minus n / x to the power minus n
$\sqrt{x}$	root x / square root x / the square root of x
$\sqrt[3]{x}$	cube root x
$\sqrt[4]{x}$	fourth root x
$\sqrt[n]{x}$	nth root x /'enθ ruːt 'eks/
$(x+y)^2$	x plus y all squared
$\left(\dfrac{x}{y}\right)^2$	x over y all squared
n!	n factorial / factorial n
$x\%$	x per cent /'eks pə 'sent/
∞	infinity
$x\propto y$	x varies as y / x is (directly) proportional to y
à	a dot /'eɪ dɒt/
ä	a double dot /'eɪ 'dʌbl dɒt/
f(x)	fx /f of x/ the function of x
f'(x)	f dash x / the (first) derivative of f with respect to x
f''(x)	f double-dash x / the second derivative of f with respect to x

$f'''(x)$	f triple-dash x / f treble-dash x / the third derivative of f with respect to x
$f^{(4)}(x)$	f four x / the fourth derivative of f with respect to x
$\dfrac{\partial v}{\partial \theta}$	the partial derivative of v with respect to θ
$\dfrac{\partial^2 v}{\partial \theta^2}$	d two v by d theta squared / the second partial derivative of v with respect to θ
w.r.t.	with respect to
$\displaystyle\int_0^\infty$	the integral from zero to infinity
$\displaystyle\sum_{i=1}^{n}$	the sum from i equals one to n
$\displaystyle\lim_{\Delta x \to 0}$	the limit as delta x approaches zero
$\displaystyle\mathop{Lt}_{\Delta x \to 0}$	the limit as delta x tends to zero
grad	gradient
div	divergence
$\log_e y$	log y to the base e / log to the base e of y / natural log (of) y
$l_n\, y$	log y to the base e / log to the base e of y / natural log (of) y
$\overrightarrow{OA}$ or $\overrightarrow{OA}$	OA / vector OA
$x \in A$	x belongs to A / x is a member of A / x is an element of A
$x \notin A$	x does not belong to A / x is not a member of A / x is not an element of A
$A \subset B$	A is contained in B / A is a proper subset of B
$A \subseteq B$	A is contained in B / A is a subset of B
$B \cap A$	B intersection A
$B \cup A$	B union A
$\cos x$	/kɒs eks/
$\sin x$	/saɪn eks/
$\tan x$	/tæn eks/
$\sec x$	/sek eks/
$\operatorname{cosec} x$	/ˈkəʊsek eks/
$\sinh x$	/ʃaɪn eks/ or /sɪntʃ eks/
$\cosh x$	/kɒʃ eks/
$\tanh x$	/θæn eks/ or /tæntʃ eks/
m_a	ma / m subscript a / m suffix a
$x_1 + x_2 + x_3 \ldots\ldots$	(usually) x one plus x two plus x three, etc.
$\lvert x \rvert$	mod x / modulus x

Appendix III

Units

S.I. Units

S.I. Units (S.I. = système international) have now been adopted by most countries of the world and should always be used. Points to note are:

a) There is no full stop after abbreviations (except, of course, at the end of a sentence).

   eg *10 cm long* NOT *10 cm. long*

b) The abbreviations do not change in the plural.

   eg *10 cm* NOT *10 cms*

   When written in full, however, the units *do* take the plural -s, (eg 10 centimetres) except when used as adjectives (eg a ten-centimetre line).

c) No unit is written with a capital letter, even when its abbreviation is written with a capital letter.

   eg *1 newton* NOT *1 Newton*

d) Note the preferred spellings:

   gramme rather than *gram*

   metre rather than *meter*

e) Numbers with more than three digits are separated into groups of three with *spaces*, not commas or full-stops.

   eg *one million* is written *1 000 000*

   NOT *1,000,000* and NOT *1.000.000*

f) Figures after the decimal point are separated in the same way.

   eg 2·732 981 326

g) The decimal point is written as a point and not as a comma.

   eg *3·141* NOT *3,141* (three point one four one)

The following are the more important S.I. units for mathematics:

Quantity	Unit	Pronunciation	Symbol
length	metre	/ˈmiːtəʳ/	m̩
mass	kilogramme	/ˈkɪləgræm/	kg
time	second	/ˈsekənd/	s
temperature	kelvin*	/ˈkelvɪn/	K
plane angle	radian	/ˈreɪdɪən/	rad
solid angle	steradian	/steˈreɪdɪən/	sr
area	square metre	/ˌskweə ˈmiːtəʳ/	m^2
volume	cubic metre	/ˌkjuːbɪk ˈmiːtəʳ/	m^3
speed	metre per second	/ˈmiːtə pə ˈsekənd/	$m\,s^{-1}$
acceleration	metre per second per second	/ˈmiːtə pə ˈsekənd pə ˈsekənd/	$m\,s^{-2}$
density	kilogramme per cubic metre	/ˈkɪləgræm pə ˌkjuːbɪk ˈmiːtə/	$kg\,m^{-3}$
force	newton**	/ˈnjuːtən/	N
pressure	newton per square metre	/ˈnjuːtən pə ˌskweə ˈmiːtə/	$N\,m^{-2}$
energy	joule	/dʒuːl/	J

Notes: *°C = °K − 273·15. The degree Celsius will continue to be widely used.

**one newton = one kilogramme metre per second per second ($1N = 1\,kg\,m\,s^{-2}$)

In addition, the following prefixes may be added to the basic units to form multiples or fractions of the values of the units:

eg $1\,000\,000\,m = 1\,Mm$ (one megametre).

Multi-plied by	Prefix	Pronun-ciation	Symbol	Multi-plied by	Prefix	Pronun-ciation	Symbol
10	deka-	/'dekə/	da	10^{-1}	deci-	/'desɪ/	d
10^2	hecto-	/'hektə/	h	10^{-2}	centi-	/'sentɪ/	c
10^3	kilo-	/'kɪlə/	k	10^{-3}	milli-	/'mɪlɪ/	m
10^6	mega-	/'megə/	M	10^{-6}	micro-	/'maɪkrə/	μ
10^9	giga-	/'gɪgə/	G	10^{-9}	nano-	/'nænə/	n
10^{12}	tera-	/'terə/	T	10^{-12}	pico-	/'pɪkə/	p
				10^{-15}	femto-	/'femtə/	f
				10^{-18}	atto-	/'ætə/	a

Other Units

Older books from Great Britain and the United States may still have non-S.I. units. Here is a list of the more important.

Unit	Pronunciation	S.I. Equivalent
inch	/ɪntʃ/	$2·54\,cm$
foot	/fʊt/	$0·30\,m$
yard	/jɑːd/	$0·91\,m$
mile	/maɪl/	$1·609\,km$
pound	/paʊnd/	$0·45\,kg$
ounce	/aʊns/	$28·35\,g$
pint	/paɪnt/	$0·57\,dm^3$
gallon	/'gælən/	$4·55\,dm^3$

Appendix IV

Irregular plurals

radius, radii /ˈreɪdɪaɪ/
nucleus, nuclei
locus, loci

axis, axes /ˈæksiːz/
analysis, analyses

basis, bases
ellipsis, ellipses
hypothesis, hypotheses
parenthesis, parentheses

matrix, matrices /ˈmeɪtrɪsiːz/ *or* matrixes /ˈmeɪtrɪksɪz/
index, indices *or* indexes
appendix, appendices *or* appendixes
vertex, vertices
apex, apices *or* apexes

polyhedron, polyhedra /ˌpɒlɪˈhiːdrə/ *or* polyhedrons /ˌpɒlɪˈhiːdrənz/
octahedron, octahedra *or* octahedrons
dodecahedron, dodecahedra *or* dodecahedrons
criterion, criteria
phenomenon, phenomena

formula, formulae /ˈfɔːmjəliː/ *or* formulas /ˈfɔːmjələz/
abscissa, abscissae *or* abscissas

frustum, frusta /ˈfrʌstə/ *or* frustums /ˈfrʌstəmz/
maximum, maxima
minimum, minima

Glossary

This list gives the pronunciation of the technical and semi-technical words used in this book and definitions of those words that are not fully explained in the text or diagrams. The first number after each entry indicates the unit in which the word first appears and the second number indicates the paragraph. Thus, *2.3* means *Unit 2, paragraph 3*.

Pronunciations are shown in the system that is used in the Longman *Dictionary of Contemporary English*. The symbols are shown in this table, with a key word for each. The letters printed in **bold type** represent the sound value of the symbol.

Consonants

p	**p**ea	f	**f**ew	ʃ	**fish**ing	h	**h**ot
b	**b**ay	v	**v**iew	ʒ	plea**s**ure	m	su**m**
t	**t**ea	θ	**th**ing	tʃ	**ch**oose	n	su**n**
d	**d**ay	ð	**th**en	dʒ	**j**ump	ŋ	su**ng**
k	**k**ey	s	**s**oon	l	**l**ed	j	**y**et
g	**g**ay	z	**z**oo	r	**r**ed	w	**w**et

Vowels

iː	sh**ee**p	ɔː	c**augh**t	eɪ	m**a**ke	ɪə	h**ere**
ɪ	sh**i**p	ʊ	p**u**t	əʊ	n**o**te	eə	th**ere**
e	b**e**d	uː	b**oo**t	aɪ	b**i**te	ʊə	p**oor**
æ	b**a**d	ʌ	c**u**t	aʊ	n**ow**	eɪə	pl**ayer**
ɑː	c**a**lm	ɜː	b**ir**d	ɔɪ	b**oy**	əʊə	l**ower**
ɒ	c**o**t	ə	**a**bout			aɪə	t**ire**
						aʊə	t**ower**
						ɔɪə	empl**oyer**

Notes

1. A small raised /ʳ/ at the end of a word means that the /r/ is pronounced if a vowel follows (at the beginning of the next word), but not otherwise. For example, *far* /fɑːʳ/ means that *far away* is pronounced /fɑːr əweɪ/ but *far down* is /fɑː daʊn/.

2. The italic /ə/ means that the sound /ə/ can be used but is often omitted. It may be found before the consonants /m, n, ŋ, l, r/ in certain positions. For example, *travel* /ˈtrævəl/ means that the pronunciation /ˈtrævəl/ is possible but /ˈtrævl/ may be more common.

3. The mark /ˈ/ means that the following syllable has *main stress*, and /ˌ/ means that the following syllable has *secondary stress*. For example, *understand* /ˌʌndəˈstænd/.

above /ə'bʌv/ 2.7

abscissa /əb'sɪsə/ horizontal line drawn from any position in the Cartesian co-ordinate system to the y-axis 3.12

absolute /'æbsəluːt/ perfect, as in absolute zero = −273·13°C = 0°K

absolute error the difference between the true (or absolute) value and the approximate value 7.10

acceleration /ək‚seləˈreɪʃən/ the rate of increase of speed 4.14

accurate /'ækjərət/ correct, exact 3.11

Achilles /əˈkɪliːz/ B.2

acute /əˈkjuːt/ (of angles) less than 90° 2.10

addition /əˈdɪʃən/ 3.2

adjacent /əˈdʒeɪsənt/ next to. adjoining 2.2 (in a right-angled triangle) the side between the angle concerned and the right angle 2.5

age /eɪdʒ/ 4.6

allow /əˈlaʊ/ 5.8

alternate /ɔːlˈtɜːnət/ one in every two, following in turn **alternate angles** 2.2

alternative /ɔːlˈtɜːnətɪv/ something which may be done instead of another 11.3

although /ɔːlˈðəʊ/ 10.6

altitude /'æltɪtjuːd/ height 1.10

anti-clockwise /‚æntɪˈklɒkwaɪz/ 11.3

apex /'eɪpeks/ the top point A.1

approach /əˈprəʊtʃ/ to get nearer and nearer to, to tend to 6.1

approximately /əˈprɒksɪmətlɪ/ 1.4

arbitrary /'ɑːbɪtrərɪ/ chosen at random, unsystematic 7.2

arc /ɑːk/ part of the circumference of a circle 1.15

area /'eərɪə/ 1.12

arithmetic /əˈrɪθmətɪk/ A.5

arithmetical /‚ærɪθˈmetɪkəl/ 4.8

arrange /əˈreɪndʒ/ 3.1

ascending /əˈsendɪŋ/ increasing, going up 4.8

auxiliary /ɔːgˈzɪljərɪ/ helping, supporting, not of central importance A.5

average /'ævrɪdʒ/ 4.8

axiom /'æksɪəm/ statement generally accepted as true without being proved 11.4

axiomatic /‚æksɪəˈmætɪk/ 11.4

axis /'æksɪs/ any line that divides a regular shape into two equal parts; (of a graph or co-ordinate system) one of the two lines on which the scale is marked 3.12

base /beɪs/ (plane geometry) the line on

which a figure stands; (solid geometry) the face on which the figure stands, or, in prisms, either of the two end faces 1.10

below /bɪˈləʊ/ 2.7

bisect /baɪˈsekt/ to cut into two equal parts 1.12

body /'bɒdɪ/ A.3

Bolyai /bɒlˈjæj/ 12.3

Boole /buːl/ 12.3

bound /baʊnd/ without bound = without limit 6.4

braces /'breɪsɪs/ 5.3

calculate /'kælkjəleɪt/ work out, compute, find the value of 4.3

calculator /'kælkjəleɪtəʳ/ 3.1

cancel /'kænsəl/ (of a fraction) to reduce by dividing numerator and denominator by the same number 3.7

capacity /kəˈpæsɪtɪ/ the amount that something can hold 4.6

capital /'kæpɪtəl/ 1.3

Cartesian /kɑːˈtiːzjən/ 3.12

case /keɪs/ 3.1

Cauchy /'kəʊʃiː/ 12.3

centre /'sentəʳ/ 1.15

centre of gravity /'sentər əv 'grævɪtɪ/ the point in anything on which it would balance C.5

centroid /'sentrɔɪd/ centre of gravity C.5

chances /'tʃɑːnsɪz/ odds, probability 10.3

chord /kɔːd/ straight line drawn between two parts of a circle or other curve 1.15

circle /'sɜːkəl/ closed curve where all points on the circumference are equidistant from the centre 1.9

circuit /'sɜːkɪt/ a path or route (e.g. a circle) which finishes at its starting-point 3.1

circular /'sɜːkjələʳ/ 1.9

circumference /səˈkʌmfərəns/ the line which forms a circle, or its length 1.15

circumscribe /'sɜːkəmskraɪb/ 2.6

clear /klɪəʳ/ 3.1

clockwise /'klɒkwaɪz/ 11.3

coin /kɔɪn/ 10.1

collide /kəˈlaɪd/ (of two or more things) to strike each other B.3

column /'kɒləm/ (in matrices) numbers (etc.) arranged vertically one under the other 2.8

common /'kɒmən/ 3.10

complement /'kɒmplɪmənt/ 5.4

complete /kəmˈpliːt/ 8.4

complex /'kɒmpleks/ not simple, e.g. complex number 3.15

comprise /kəm'praɪz/ consist of 3.1

computer /kəm'pjuːtəʳ/ A.5

concurrent /kən'kʌrənt/ meeting at one point C.5

cone /kəʊn/ solid figure formed by rotating a right-angled triangle round one side (not the hypotenuse) through 360° 4.2

congruent /'kɒngrʊənt/ 1.10

consecutive /kən'sekjətɪv/ directly following each other 10.4

consequently /'kɒnsɪkwəntlɪ/ as a result 6.4

considerable /kən'sɪdərəbəl/ 7.3

consist of /kən'sɪst əv/ 3.1

constant/'kɒnstənt/ invariable, unchanging (usually denoted in equations by letters at the beginning of the alphabet) 8.2

contain /kən'teɪn/ 3.1

contradiction /,kɒntrə'dɪkʃən/ 11.6

control /kən'trəʊl/ A.5

converge /kən'vɜːdʒ/ to come together, to approach a point 6.4

convergent /kən'vɜːdʒənt/ getting nearer and nearer to some point 6.4

co-ordinate /kəʊ'ɔːdənət/ 3.12

correct /kə'rekt/ 3.11

correspond /,kɒrɪs'pɒnd/ to match, be in a comparable position to 2.2

corresponding /,kɒrɪs'pɒndɪŋ/ 2.2

cosecant /'kəʊsiːkənt/ hypotenuse divided by opposite 6.3

cosine /'kəʊsaɪn/ adjacent divided by hypotenuse 2.5

cotangent /kəʊ'tændʒənt/ adjacent divided by opposite 6.3

cross-sectional /,krɒs'sekʃənəl/ 4.1

cube /kjuːb/ regular polyhedron with square faces 4.4

current /'kʌrənt/ flow of electricity 5.8

cursor /'kɜːsəʳ/ A.3

curve /kɜːv/ 1.1

curved /kɜːvd/ 1.1

cylindrical /sɪ'lɪndrɪkəl/ 4.3

decagon /'dekəgən/ ten-sided figure 1.5

decimal point /'desɪməl 'pɔɪnt/ 3.1

decrease /dɪ'kriːs/ to become smaller or less 6.3

deduction /dɪ'dʌkʃən/ 11.6

deep /diːp/ 4.6

degree /dɪ'griː/ 2.10

denominator /dɪ'nɒmɪneɪtəʳ/ in a fraction) the part below the line 3.6

denote /dɪ'nəʊt/ 5.3

density /'densɪtɪ/ 4.6

depth /depθ/ 4.6

Descartes /deɪ'kɑːt/ 12.3

descending /dɪ'sendɪŋ/ decreasing, going down 4.8

determine /dɪ'tɜːmɪn/ to find the exact position of 8.9

device /dɪ'vaɪs/ 5.1

diagonal /daɪ'ægənəl/ a straight line joining one vertex of a geometrical figure to another 1.12

diagrammatically /,daɪəgrə'mætɪklɪ/ by using a diagram 5.1

diameter /daɪ'æmɪtə/ a straight line joining two points on the circumference of a circle and passing through its centre 1.15

dice /daɪs/ 10.1

difference /'dɪfərəns/ the result of subtracting one number from another 3.2

digit /'dɪdʒɪt/ any of the ten numbers from 0 to 9 3.1

direct /dɪ'rekt/ 9.5

disjoint /dɪs'dʒɔɪnt/ not touching or intersecting; disjoint sets 5.4

displacement /dɪs'pleɪsmənt/ distance where both magnitude and direction are given 4.14

display /dɪs'pleɪ/ 3.1

distance /'dɪstəns/ 4.14

divergent /daɪ'vɜːdʒənt, dɪ'vɜːdʒənt/ not convergent 6.4

diverge /daɪ'vɜːdʒ, dɪ'vɜːdʒ/ to move apart, to move away from a point 6.4

divisible /dɪ'vɪzɪbəl/ 3.3

division /dɪ'vɪʒən/ 1.4

dodecahedron /,dəʊdekə'hiːdrən/ regular polyhedron with twelve pentagonal faces 4.4

duration /djʊ'reɪʃən/ length of time B.3

edge /edʒ/ 4.4

effect /ɪ'fekt/ 8.12

elasticity /iːlæ'stɪsɪtɪ/ ability to regain original shape after distortion B.4

element /'elɪmənt/ 2.8

elimination /ɪ,lɪmɪ'neɪʃən/ 11.6

eliminate /ɪ'lɪmɪneɪt/ remove, get rid of C.2

empty /'emptɪ/ 5.3

enable /ɪ'neɪbəl/ 5.2

equal /'iːkwəl/ 1.4

equation /ɪ'kweɪʒən/ 7.4

equidistant /,iːkwɪ'dɪstənt/ at an equal distance from 1.15

equilateral /ˌiːkwɪˈlætərəl/ with sides of equal length 1.6

error /ˈerəʳ/ 7.10

escribe /ɪˈskraɪb/ C.6

Euclid /ˈjuːklɪd/ 12.3

Euler /ˈjuːləʳ/ 4.5

even /ˈiːvən/ 3.3

exact /ɪgˈzækt/ 7.1

exclude /ɪkˈskluːd/ 10.2

exclusive /ɪkˈskluːsɪv/ 10.2

exert /ɪgˈzɜːt/ B.3

exhaustive /ɪgˈzɔːstɪv/ covering all the possibilities 10.2

exterior /ɪkˈstɪərɪəʳ/ 2.2

face /feɪs/ A.1

factor /ˈfæktəʳ/ 3.9

factorial /fækˈtɔːrɪəl/ 1.4

factorise /ˈfæktəraɪz/ to enumerate the factors of 3.9

Fibonacci /fɪbəˈnætʃɪ/ 12.8

figure /ˈfɪgəʳ/ 1.5

final /ˈfaɪnəl/ at the end B.3

flaw /flɔː/ logical error 5.15

force /fɔːs/ 4.14

formula /ˈfɔːmjələ/ 4.1

fraction /ˈfrækʃən/ 3.5

frustum /ˈfrʌstəm/ A.2

function /ˈfʌnkʃən/ 6.3

further /ˈfɜːðəʳ/ more 7.7

general /ˈdʒenərəl/ 8.2

generality /ˌdʒenəˈrælɪtɪ/ 12.1

geometric /ˌdʒiːəˈmetrɪk/ 1.15

gradient /ˈgreɪdɪənt/ slope 6.7

graph /græf, grɑːf/ 5.1

greatest common divisor /ˈgreɪtɪst ˈkɒmən dɪˈvaɪzəʳ/ 3.10

hardware /ˈhɑːdweəʳ/ the machinery which makes up a computer A.5

heads /hedz/ 10.1

height /haɪt/ 4.1

hence /hens/ 9.7

heptagon /ˈheptəgən/ seven-sided figure 1.5

heptagonal /hepˈtægənəl/ 1.9

hexagon /ˈheksəgən/ six-sided figure 1.5

hexagonal /hekˈsægənəl/ 1.9

hexahedron /ˌheksəˈhiːdrən/ cube 4.4

high /haɪ/ 4.6

highest common factor /ˈhaɪəst ˈkɒmən ˈfæktəʳ/ 3.10

horizontal /ˌhɒrɪˈzɒntəl/ 1.1

however /haʊˈevəʳ/ (= it does not matter which) 6.2

hypotenuse /haɪˈpɒtənjuːz/ (in a right-angled triangle) the side opposite the right angle 2.5

icosahedron /ˌaɪkɒsəˈhiːdrən/ regular polyhedron with twenty triangular faces 4.4

identical /aɪˈdentɪkəl/ exactly the same A.3

illustrate /ˈɪləstreɪt/ to give an example of 11.5

imaginary /ɪˈmædʒɪnərɪ/ 3.15

immediately /ɪˈmiːdɪətlɪ/ 2.7

impact /ˈɪmpækt/ hitting B.3

improper /ɪmˈprɒpəʳ/ (of fraction) where numerator is greater than denominator 3.7

impulse /ˈɪmpʌls/ B.3

include /ɪnˈkluːd/ 3.1

increase /ɪnˈkriːs/ to become greater or more 6.3

inequality /ˌɪnɪˈkwɒlɪtɪ/ 7.4

infinity /ɪnˈfɪnɪtɪ/ 6.3

inflexion /ɪnˈflekʃən/ (sometimes *inflection*) bending or curving 6.5

initial /ɪˈnɪʃəl/ first, at the beginning B.3

input /ˈɪnpʊt/ A.5

inscribe /ɪnˈskraɪb/ 2.6

instrument /ˈɪnstrəmənt/ 5.1

insufficient /ˌɪnsəˈfɪʃənt/ 7.5

integer /ˈɪntɪdʒəʳ/ whole number, number not involving a fraction 3.3

interior /ɪnˈtɪərɪəʳ/ 2.2

intersect /ˌɪntəˈsekt/ 1.12

intersection /ˌɪntəˈsekʃən/ (in geometry) the point where two lines meet; (in set theory) 5.4

inverse /ɪnˈvɜːs/ opposite way round; inversely proportional 9.5

involve /ɪnˈvɒlv/ to have as a part 11.10

irrational /ɪˈræʃənəl/ 3.15

isosceles /aɪˈsɒsəliːz/ (of triangle) with two sides equal (of trapezium) with the two non-parallel sides equal 1.6

joint /dʒɔɪnt/ shared by two or more things; jointly proportional 9.5

Kepler /ˈkepləʳ/ 12.4

key /kiː/ 3.1

lateral /ˈlætərəl/ at the side A.1

law /lɔː/ 11.4

Leibniz /ˈlaɪbnɪts/ 12.3

length /lenθ/ 4.1

level /ˈlevəl/ standard, degree 12.1

likely /ˈlaɪklɪ/ probable 10.3

limit /ˈlɪmɪt/ 6.4

line /laɪn/ 1.1

linear /ˈlɪnɪə/ (of an equation) of first degree; of which the graph is a straight line 11.9

Lobachevsky /ˌlɒbəˈtʃevskɪ/ 12.3

logic /'lɒdʒɪk/ 5.8

long /lɒŋ/ 4.6

loudness /'laʊdnəs/ magnitude of noise, measured in decibels 4.7

lowest common multiple /'ləʊəst 'kɒmən 'mʌltəpəl/ 3.10

magnitude /'mægnɪtjuːd/ 4.14

mass /mæs/ 4.14

matrix /'meɪtrɪks/ 2.7

maximum /'mæksɪməm/ 4.8

mean /miːn/ 4.8

measure /'meʒəʳ/ 4.8

median /'miːdɪən/ (in geometry) C.5 (kind of average) 4.8

member /'membəʳ/ 5.4

memory /'meməri/ A.5

midrange /,mɪd'reɪndʒ/ 4.8

minimum /'mɪnɪməm/ 4.8

minus /'maɪnəs/ 1.4

misleading /mɪs'liːdɪŋ/ giving a false impression 9.9

Möbius /'miːbɪəs/ 12.5

mode /məʊd/ 4.8

momentum /mə'mentəm/ 4.14

moveable /'muːvəbəl/ which can be moved A.3

multiplicand /,mʌltɪplɪ'kænd/ the number which is multiplied 7.9

multiplication /,mʌltɪplɪ'keɪʃən/ 1.4

mutually exclusive /'mjuːt ʃʊəlɪ ɪk'skluːsɪv/ 10.2

Napier /'neɪpɪəʳ/ 12.4

natural /'nætʃərəl/ 5.12

necessary /'nesəsəri/ 7.6

Newton /'njuːtən/ 12.3

notation /nə'teɪʃən/ 5.4

numerator /'njuːməreɪtəʳ/ (in a fraction) the part above the line 3.6

oblique /ə'bliːk/ 1.1

obtuse /əb'tjuːs/ (of angles) more than 90° and less than 180° 2.10

occasionally /ə'keɪʒənəlɪ/ from time to time 10.5

octagon /'ɒktəgən/ eight-sided figure 1.5

octagonal /ɒk'tægənəl/ 1.9

octahedron /,ɒktə'hiːdrən/ regular polyhedron with eight triangular faces 4.4

odd /ɒd/ 3.3

opposite /'ɒpəzɪt/ (in a right-angled triangle) the side opposite the angle concerned 2.5; 1.6

ordered /'ɔːdəd/ put into systematic order 4.8

ordinate /'ɔːdənət/ vertical line drawn from any position in the Cartesian co-ordinate system to the x-axis 3.12

origin /'ɒrɪdʒɪn/ the point of intersection of the two axes of a graph or co-ordinate system 3.12

orthocentre /'ɔːθəsentəʳ/ C.5

outcome /'aʊtkʌm/ 10.1

output /'aʊtpʊt/ A.5

parallel /'pærəlel/ 1.1

parallelogram /,pærə'leləgræm/ quadrilateral with opposite sides parallel 1.6

Pascal /pæ'skæl/ 10.9

pentagon /'pentəgən/ five-sided figure 1.5

pentagonal /pen'tægənəl/ 1.9

percentage /pə'sentɪdʒ/ 3.1

perfect /'pɜːfɪkt/ 8.4

periodic /,pɪərɪ'ɒdɪk/ repeating itself 6.3

peripheral /,pə'rɪfərəl/ not of central importance A.5

perpendicular /,pɜːpən'dɪkjʊləʳ/ (adj.) at right angles (noun) a line at right angles to another line 1.1

plane /pleɪn/ two-dimensional 1.5

plus /plʌs/ 1.4

point /pɔɪnt/ 1.1

polar /'pəʊləʳ/ 8.13

polygon /'pɒlɪgən/ n-sided plane figure 6.1

polyhedron /,pɒlɪ'hiːdrən/ solid figure with n sides 4.4

power /'paʊəʳ/ (electrical) 5.7; 4.14 (mathematical) 8.1

precede /prɪ'siːd/ to go before 4.11

pressure /'preʃəʳ/ 9.6

prevent /prɪ'vent/ 5.10

prime /praɪm/ = prime number – number with no factors except 1 and itself 3.4

prism /'prɪzəm/ solid figure with congruent ends (or bases) and with parallelograms as its lateral faces 4.1

produce /prə'djuːs/ extend 1.17

product /'prɒdʌkt/ the result of multiplying one number by another 3.2

progression /prə'greʃən/ a set of numbers which follow each other in a regular way 4.11

proof /pruːf/ 9.8

proper /'prɒpəʳ/ (of fraction) where numerator is smaller than denominator 3.7

property /'prɒpəti/ 1.1

proportion /prə'pɔːʃən/ 9.7

proportional /prə'pɔːʃənəl/ 9.5

protractor /prə'træktəʳ/ an instrument